Technical Writing for Beginners

Winston Smock

TECHNICAL WRITING FOR BEGINNERS

Prentice-Hall, Inc.,
Englewood Cliffs, New Jersey 07632

Library of Congress Cataloging in Publication Data

Smock, Winston.
Technical writing for beginners.

"A Spectrum Book."
Includes index.
1. English language—Technical English. 2. English language—Rhetoric. 3. Technical writing. 4. Technical publishing. I. Title.
PE1475.S66 1984 808'.06665 83-23040
ISBN 0-13-898452-2
ISBN 0-13-898445-X (pbk.)

A Spectrum Book. Printed in the United States of America.

1 2 3 4 5 6 7 8 9 10

ISBN 0-13-898452-2

ISBN 0-13-898445-X {PBK.}

Editorial/production supervision
and book design by Eric Newman
Cover design © 1983 by Jeannette Jacobs
Manufacturing buyer: Edward J. Ellis

This book is available at a special discount when ordered in bulk quantities. Contact Prentice-Hall, Inc., General Publishing Division, Special Sales, Englewood Cliffs, N.J. 07632.

PRENTICE-HALL INTERNATIONAL, INC., *London*
PRENTICE-HALL OF AUSTRALIA PTY. LIMITED, *Sydney*
PRENTICE-HALL CANADA INC., *Toronto*
PRENTICE-HALL OF INDIA PRIVATE LIMITED, *New Delhi*
PRENTICE-HALL OF JAPAN, INC., *Tokyo*
PRENTICE-HALL OF SOUTHEAST ASIA PTE. LTD., *Singapore*
WHITEHALL BOOKS LIMITED, *Wellington, New Zealand*
EDITORA PRENTICE-HALL DO BRASIL LTDA., *Rio de Janeiro*

for Cecilia

Contents

WRITING AND PUBLISHING

Preface

This book is intended to be an informative compilation of things that a beginning technical writer needs to know. It consists of three parts: an introduction to science and technology, a handbook of style, and a step-by-step description of the publishing process. The objectives are (1) to give the literary-oriented reader a smattering of science and technology, (2) to give the technically oriented reader some sense of literary style, and (3) to give all readers some insight into the technical writing and publishing process.

The writer who is ignorant of science and technology gives himself away in many subtle ways. For this reason, a considerable portion of this book is devoted to a discussion of selected scientific and technical topics that may prove helpful to the beginning technical writer. Part I of this book is a gentle introduction to science and technology for the benefit of those of you who are approaching the field of technical writing from a background in the liberal arts.

Technical writing is characterized by an obsession with numbers and measurement. Proficiency in the use of numbers and the symbols of measurement is indispensable to the technical writer. An entire chapter (Chapter 9) is devoted to this very important topic. Equally important, Chapter 7, "Abbreviations, Symbols, and Units," is a guide to the use of abbreviations in technical writing.

Even if armed with a Ph.D. in physical science and a master's degree in English, a beginning technical writer will have many questions to ask—and no one to ask them of. Part III of this book is designed to ease the transition from novice to professional technical writer by answering a number of these questions in advance.

Perhaps this book's most important feature is a step-by-step description of the technical publishing process, from the inception of a writing project to the delivery of printed copies. This process includes research, writing, revision, editing, production, and printing. There is a chapter on the content, style, and format of technical manuals and one on research and organization. There is a list of elements that should be included in every technical manual and a lengthy passage on copy preparation. Considerable space is given to the construction of tables and the treatment of illustrations. There is a chapter on technical review and revision, and the book concludes with a chapter on production and printing.

I want to take this opportunity to thank Titus Arbiter, Lynn Doyle, Victor Houseman, and countless others whose names I can't remember, who read and commented on parts of this work when it was still a thirty-six-page in-house style guide and whose valuable comments are an inseparable part of its fabric. Special thanks are due to my wife, Cecilia, and to Hannerl and Hermann Ebenhoech, who read the completed manuscript and gave me the courage to continue.

Technical Writing for Beginners

PART ONE
SCIENCE AND TECHNOLOGY

If you have entered the field of technical writing from a background in science or technology, there will not be much that is new to you in the following chapters; and you may proceed with profit directly to Part II. If, on the other hand, your background is light in technology, it may be to your advantage to read these chapters.

A Smattering of Science. Science and technology are natural partners. Were it not for science, we would be stuck with the technology of the dark ages. Were it not for technology, science would be limited to naked-eye observations.

What is science? In a word, science is a method. It is a way of arriving at truth without fasting or prayer, or mind-altering drugs. It is the only path to enlightenment that has produced a technology worthy of the name.

There is nothing esoteric about the scientific method; it is the way that we go about solving ordinary problems in our ordinary lives. If we want to know the answer to a question, we look and see.

Science is a method of controlled observation. It works like this: An investigator poses a question that he or she wants answered, conceives of an observation or experiment that might yield an answer or part of an answer, performs the necessary observation or experiment, records the observations in sufficient detail to allow others to repeat his or her work, and publishes the results in a learned journal. Other investigators repeat the experiment or observations and either confirm or refute the findings. In this way, the frontiers of knowledge are advanced bit by painful bit. The method cannot compare with revelation for speed and ease, but it makes up for this shortcoming in validity and reliability.

In the following chapters, you will find a smattering of physical science, things that patient investigators have found out by the painful process outlined in the foregoing paragraphs.

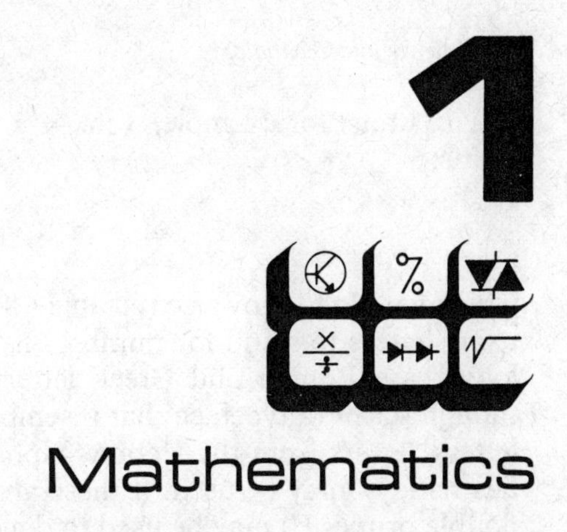

1 Mathematics

The producers and consumers of science feel confidence in observations and conclusions that are expressed quantitatively. For this reason, the literature of science and technology is heavily larded with numbers and the signs and symbols of mathematics. If your eyes tend to glaze over at the sight of an equation, you are not alone. The aim of this chapter is to give you, ever so gently, an inkling of what mathematics is all about—no more.

Although there are many branches of mathematics, technical writing relies rather heavily on three disciplines: algebra, trigonometry, and the calculus. Algebra is a powerful mathematical language that allows the scientist to express scientific generalizations in very succinct form. $E = mc^2$ is an algebraic expression. Trigonometry, originally developed to deal with the measurement of triangles, is now much used in describing repetitive or cyclic phenomena. The calculus is a powerful tool useful for describing nonrepetitive changes.

□ ALGEBRA

To find the area of a rectangle that measures 3 units by 5 units, one multiplies the length by the width and expresses the area in square units; for example, 3 in. × 5 in. = 15 sq in. In general terms:

$$A = l \times w,$$

which will do for any size rectangle. This trick of generalizing by using letters to stand for numbers is called algebra. Upper- and lower-case Roman and Greek letters are commonly used, although script (a typeface that resembles longhand) and German letters, letters from the Hebrew alphabet—whatever the printer has handy—may be used. If these aren't enough, primes (′) and double primes (″) may be used to double or triple the number of algebraic quantities at our disposal. Letters from the first part of the alphabet (a, b, c) are commonly used to represent known quantities, and letters from the end of the alphabet (x, y, z) to represent unknown quantities.

Some letters have become so firmly identified with certain quantities or classes of quantities as to have become virtually unavailable for general use. Among these are ϕ and Θ to represent angles, π to represent the ratio of the diameter to the circumference of a circle, ω to represent angular frequency, i to represent the square root of -1, and e to represent the base of the natural system of logarithms.

Subscripts. Subscripts may be used to increase our arsenal of symbols without limit. For instance, $t_0, t_1, t_2, \ldots t_n$ may represent a series of points along a time scale.

Superscripts. Superscripts (except for the primes mentioned previously) perform a function that is quite different from that performed by subscripts. A superscript indicates that a quantity is to be multiplied by itself. A^2 (read "A squared") means $A \times A$. A^3 (read "A cubed") means $A \times A \times A$. A^7 means seven A's multiplied together. Superscripts used in this manner are called exponents. Exponents may be positive, negative, or fractional. Negative exponents mean that the preceding quantity is to be divided by the indicated power of the quantity bearing the exponent ($a \times b^{-2} = a/b^2$).

Scientific Notation. Very large and very small numbers are conveniently presented as a number between 1 and 10, multiplied by a power of 10. (This saves writing out all those zeroes and possibly miscounting them, which could result in a very big error.) Thus, 5,000 can be written as 5×10^3 and 0.0000005 can be written as 5×10^{-7}.

Fractional Exponents. An exponent does not have to be a whole number. $A^{1/2}$ equals the square root of A, and $A^{1/3}$ equals the cube root of A. Mixed numbers, too, may be used as exponents; for instance: $10^{2.5}$ equals some number between 100 and 1000 (316.228, actually).

Exponential Equations. The exponent, too, may be an algebraic quantity. Equations involving algebraic powers of numbers (for example, a^x) are known as "exponential" equations.

Logarithms. If a equals 10^x, then x is said to be the *logarithm* of a to the base 10 (written "log a"). Logarithms to the base 10 were widely used for computation prior to the advent of the pocket calculator but are not much used any more. Another base, e (= 2.7183), is still quite commonly used. The reason for the choice of this irrational number is that certain calculus expressions are thereby somewhat simplified. Logarithms to the base e are known as "natural" logarithms, abbreviated ln.

Operators. In addition to letters, algebra also uses symbols called "operators"; for example, $+$, $-$, $\times$, $/$, and $=$, which indicate the familiar arithmetic operations of addition, subtraction, multiplication, and division, and the property of equality, respectively.

In addition to the "times" sign ($\times$), multiplication may be indicated by a raised "product dot ($\cdot$)," or by mere juxtaposition. $Ab = A \times b$. Other operators sometimes seen are Δ, which means "change" or "difference," and Σ, which means "sum." Increasingly, you may expect to see an asterisk ($*$) used in place of the $\times$ sign. This usage comes from computer programming. Computer terminals do not have a "times" sign that is different from the letter *X*.

Grouping. In complex expressions, the letters and symbols are grouped within parentheses or under a fraction bar to indicate the order in which the operations are to be performed; $(a + b) \times c$ does not mean the same as $a + (b \times c)$.

□ TRIGONOMETRY

Our hairy ancestors knew how to count long before they learned how to write. One of the things that they liked to count was the days of the year, and the figure that they came up with was 360. (They didn't exactly ignore the five extra days; they had a party. They also knew how to make beer.) Three hundred and sixty is a splendid number; it is divisible by every integer from 1 to 10, except 7 (thus an unfortunate choice for the number of days in the week).

For a long time, the apparent motions of the sun and planets and the female menstrual cycle (which they got mixed up with the phases of the moon) were the only cyclic phenomena known. (This was before the invention of the wheel.) Later on, when other cyclic phenomena came to be investigated, "360 degrees" was used as a synonym for a full cycle, "180 degrees" for a half cycle, and so on. Thus, the concept of angular measurement was born.

Simplified Definitions (without Triangles). Just so that the trigonometric functions will not be a complete mystery, the following simplified discussion is offered.

Consider a point P on the circumference of a circle of unit radius (Figure 1). The vertical distance y of P from the horizontal axis is known as the sine of angle Θ ($\sin \Theta$), which can be either positive or negative. The horizontal distance of P from the vertical axis is the cosine of Θ ($\cos \Theta$). The ratio of the sine to the cosine (y/x) is the tangent ($\tan \Theta$); and its inverse, x/y, is the cotangent ($\cot \Theta$). The reciprocal of the cosine ($1/x$) is the secant ($\sec \Theta$), and the reciprocal of the sine ($1/y$) is the cosecant ($\csc \Theta$). The foregoing simplified definitions are good for a circle of unit radius only. For any other value of r, the definitions are:

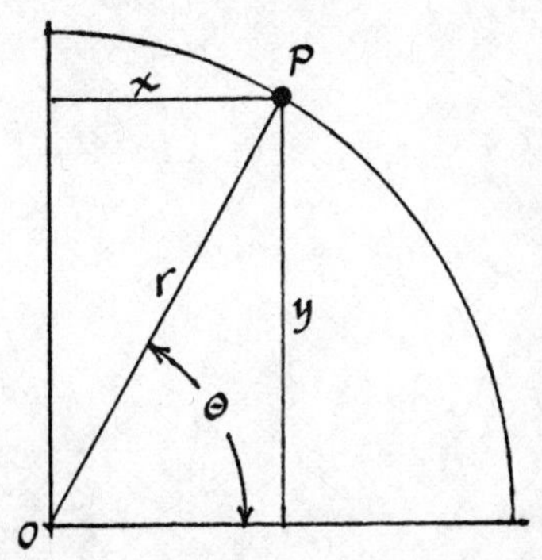

Figure 1. Trigonometry without triangles. Consider a point P on the circumference of a circle of unit radius. The distance y of that point from the horizontal axis is the sine of the angle θ. The distance x of the point from the vertical axis is the cosine of θ. The ratio y/x is the tangent of θ, and its inverse, x/y, is the cotangent of θ. The reciprocal of the cosine ($1/x$) is the secant, and the reciprocal of the sine ($1/y$) is the cosecant. These definitions hold for a circle of unit radius only. See text for more generalized definitions.

$$\begin{aligned}
\sin\Theta &= y/r, \\
\cos\Theta &= x/r, \\
\tan\Theta &= y/x, \\
\cot\Theta &= x/y, \\
\sec\Theta &= r/x, \text{ and} \\
\csc\Theta &= r/y.
\end{aligned}$$

These six trigonometric functions are much used in science in describing cyclic phenomena. A graph of the sine function is shown in Figure 2.

□ THE CALCULUS

Just as algebra is useful in the study of static measurements and trigonometry in the study of cyclic phenomena, so the calculus is useful in the study of phenomena that are neither static nor repetitive but that change in other ways.

Like trigonometry, the calculus is concerned with ratios, seeking to describe nonrepetitive phenomena in terms of the ratios of infinitesimal quantities. (If the concept of an infinitesimal quantity is philosophically repugnant to you, consider that no matter how small a quantity you name, I can name a smaller

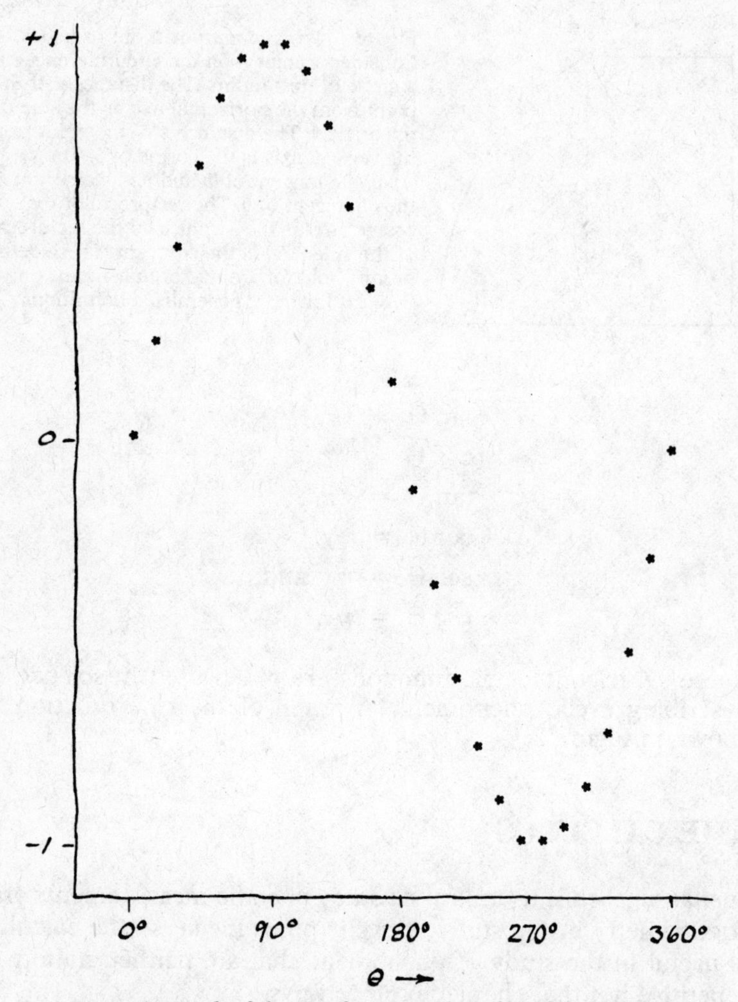

Figure 2. Graph of the sine function. The value of the sine is 0 at 0°, 180°, and 360°. It is +1 at 90° and −1 at 270°.

one.) The ratios can be quite large, even when the quantities are very small.

Typically, the ratios are expressed in the form

$$dy/dx$$

where the letter *d*, though italic, is not an algebraic quantity but an operator like +, −, ×, and / and means "take an infinitesimal bit of" the following value. Thus, *dy* means "take an infinitesimal bit of *y*," and the expression *dy/dx* expresses the effect on *y* of an infinitesimal change in *x* and vice versa. The process is called differentiation.

Integration. Integration, symbolized by the long ∫, is the reverse of differentiation; and the expression

$$\int dx$$

means "add up all the infinitesimal bits of *x*." (The result of this operation should be *x*. The fact that it usually is not shouldn't concern us here.)

Partial Derivatives. Sometimes, a value may depend on more than one variable. The way that mathematicians handle this is to treat one or more variables as constants while they vary another. The process is known as partial differentiation and is symbolized as δx, where the backwards six (pronounced "delta") means about the same as the operator *d* but serves to remind the reader that the indicated operation does not tell the whole story.

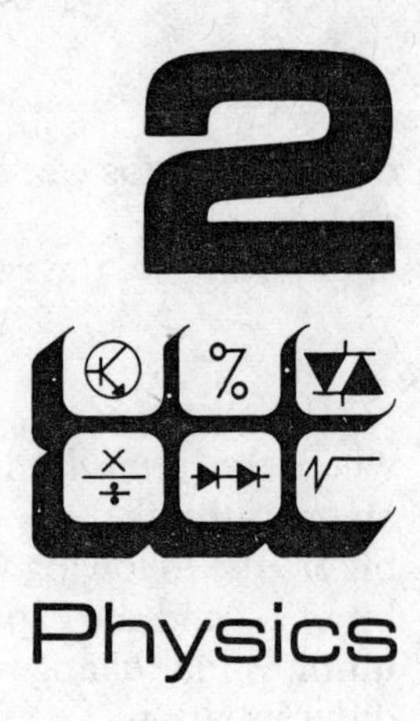

2 Physics

The science of physics embraces many disciplines. Most are beyond the scope of this work. A few (hydraulics, pneumatics, heat, electricity) have special significance for the technical writer because of their prominence in industry.

□ THE PROPERTIES OF MATTER

Central to the study of physics are the properties of matter. Some of these properties are discussed in the following paragraphs.

Matter and Energy. The name of Albert Einstein is likely to appear on anyone's list of the greatest intellects of all time. His elegant formula, $E = mc^2$, even if we don't know what it means, is certainly familiar to all of us. What it means is simply that matter and energy, although traditionally measured in different units, are really the same thing. This proposition, which seems so counter to what we intuitively know about matter and energy, was abundantly confirmed in large-scale experiments at Hiroshima and Nagasaki. If we want to know the amount of energy contained in a material object, we have merely to multiply its mass by the very large number obtained by squaring the speed of light ($c^2 = 9 \times 10^{16}$). In principle, this is no different from determining the value in pesos of x number of dollars if you know the exchange rate.

Energy is the basic stuff of the universe. This sounds almost metaphysical, but it isn't. Hard-headed physicists are in solid agreement on this point. Energy is manifest in various ways, depending upon how you look at it. Radiant energy, which includes visible light, heat, radio waves, and X rays, is described by the following equation:

$$E = h\nu$$

where E stands for the amount of energy, h is a constant multiplier ("Planck's constant" $= 6.625 \times 10^{-34}$), and ν is the frequency of the radiation. (We shall return to this point later.)

Mass. The term *mass* refers to the amount of matter that is present in a material body. All material bodies have mass. The mass of one body may be compared with that of another by comparing the force by which they are accelerated in a gravitational field, this force being called weight. The acceleration of gravity is fairly constant on the surface of the Earth, and the weight of a body is frequently used as a synonym for its mass. It must be remembered, however, that the mass of a one-kilogram object will still be one kilogram on the surface of the moon, but its weight will be much less.

Force. *Force* may be defined as any influence tending to produce a change in the motion of a material object. A body moving in a circle or following a curved trajectory is continuously changing direction. This does not happen without the application of force. Uniform motion in a straight line does not involve force. Four different forces are recognized in physics. Two of them, the so-called "strong" and "weak" forces, operate at subatomic distances (distances that are small by comparison with the diameter of an atom) and are of little concern to the technical writer. The other two, gravitation and electromagnetism, operate over vast (i.e., intergalactic) distances.

Every material body in the universe attracts every other body with a force that is directly proportional to its mass and inversely proportional to the square of the distance separating the two bodies. This is Newton's law of gravitation.

Volume. A characteristic of material bodies is that they occupy space. The ratio of mass to volume is called density.

Position. On the macrocosmic scale, one of the things that can be said about a material body is that it is somewhere. Its position is always relative to that of some other body, which is to say that if there were only one body in space, it would have no position.

Motion. Another property of matter is motion, or change of position. Like position, motion is relative. If there were only one body in space, it would have no motion.

A material body will continue its proper motion in a straight line unless it is acted upon by some external force. This is Newton's First Law of motion. Any change in the motion of a body—either in speed or direction—is called acceleration. Acceleration is not a natural property of matter and hence requires the application of force.

The acceleration that we are familiar with is positive acceleration. If we are deprived of visual clues, we cannot distinguish between positive acceleration, negative acceleration or "deceleration," acceleration due to a continuous change of direction (as in a centrifuge), or the acceleration due to gravity; it all feels the same. Blindfolded, we would be hard put to distinguish one from another.

Temperature. The universe as perceived by the naked eye is the macrocosm. The world of atoms and electrons, imperceptible to the naked eye, is the microcosm. On the microcosmic scale, it is difficult to determine the positions and motions of individual atoms and subatomic particles. Indeed, Heisenberg's Uncertainty

Principle states that you can know one only if the other is unknown; that is, you may know the position or the velocity of a particle but not both. You can, however, measure the average acceleration of the molecules in a body; and that average acceleration is called temperature. Temperature is absolute. What we perceive as temperature is the effect of the random motion of molecules. The greater the average motion the warmer the temperature.

Radiation. The molecules that constitute a material body are in constant, random motion. Restrained by the cohesive forces that hold a material body together, the molecules oscillate about fixed positions relative to one another; and a solid body tends to retain its size and shape.

Oscillating molecules radiate energy into space; and this radiant energy, appropriately enough, is called radiation. Light, heat, radio and television waves, X rays, and various other manifestations of radiant energy are examples of radiation. Radiation, like mass and temperature, is a property of matter. Radiation is characterized by its intensity and frequency.

We have an intuitive grasp of what is meant by "matter," but what is frequency? If a city bus passes a given point every twenty minutes, we would be correct in assigning to that event a frequency of three times per hour. We thus have an intuitive grasp of what is meant by "frequency"; it is the number of times that a given event occurs in a unit of time.

Buses, of course, are notoriously slow; and an hour is an intolerably long period. Atomic events happen thousands, millions, or billions of times in a second, but the principle is the same. The frequency of radiation refers to the number of oscillations per second.

How can frequency be a manifestation of energy? Think of anything that you do at regular intervals. Now imagine doing it twice as often, ten times, a hundred times as often. Do you get the idea? As the frequency goes up, so does the amount of energy involved. And since matter and energy are equivalent (per Einstein's equation), frequency is seen to be a property of matter.

The energy associated with high-frequency events doesn't accumulate in one place; it is radiated into space. Light, heat, radio emanations, and many other phenomena are examples of radiation.

Electromagnetic radiation travels at the speed of light: 3×10^8 meters per second. Since the speed is constant, the frequency can also be expressed in terms of the distance from one wave crest to the next. To return to our bus analogy, if we know that the buses travel at a steady twenty miles per hour, it is easy to calculate the distance between buses if we know the frequency, and vice versa. Thus, if there are three buses per hour and they travel at twenty miles per hour, the average distance between buses must be 20/3 = 6.66 miles.

Radiation, like motion, force, mass, and temperature, is a property of matter. Any material body at a temperature greater than absolute zero (zero kelvin) radiates energy.

Radiant energy is characterized by its frequency or wavelength. The terms *frequency* and *wavelength* represent two different ways of talking about the same thing. Frequency and wavelength are related by the formula

$$f\lambda = 3 \times 10^8$$

where f = the frequency in hertz (cycles per second),
λ = the wavelength in meters, and
3×10^8 = the speed of light in meters per second.

Hence, it may be seen that the higher the frequency, the shorter the wavelength; the longer the wavelength, the lower the frequency. If you want to know the wavelength of a given frequency, use the formula

$$\lambda = \frac{3 \times 10^8}{f}$$

If you want to know the frequency of a given wavelength, use the formula

$$f = \frac{3 \times 10^8}{\lambda}$$

Always remember: long waves = low frequencies; short waves = high frequencies.

Hence, for every frequency there is a specific wavelength, and vice versa (see Table 1). High temperatures produce high-frequency radiation.

Radiant energy is loosely spoken of as "light," although visible light occupies only a very narrow portion of the electromagnetic spectrum. Wavelengths somewhat longer than those of visible light are perceived as heat. Shorter wavelengths are not perceived directly but are detectable by other means (i.e., instruments).

Table 1. The Electromagnetic Spectrum.

Typical radiation	*Frequency (Hz)*	*Wavelength (m)*
Communications	10^4	3×10^4
	10^5	3×10^3
	10^6	3×10^2
Television (VHF)	10^7	3×10^1
Television (UHF)	10^8	3×10^0
	10^9	3×10^{-1}
	10^{10}	3×10^{-2}
	10^{11}	3×10^{-3}
Infrared	10^{12}	3×10^{-4}
Infrared	10^{13}	3×10^{-5}
Visible light[a]	10^{14}	3×10^{-6}
Ultraviolet	10^{15}	3×10^{-7}
Ultraviolet	10^{16}	3×10^{-8}
X rays	10^{17}	3×10^{-9}
X rays	10^{18}	3×10^{-10}
X rays	10^{19}	3×10^{-11}
Gamma rays	10^{20}	3×10^{-12}

[a]Visible light occupies a narrow band of frequencies between 4 and 8×10^{14} Hz.

We perceive radiation (and gravitation) in terms of its local effect. Although the amount of radiation emanating from a body is constant regardless of distance (that is, radiation does not diminish with time nor with distance traveled), a given amount of radiation is spread over a greater area at a greater distance from the source. This is expressed as the "inverse square law," which states that the intensity of radiation varies inversely as the square of the distance from the source; that is: At twice the distance, the intensity is one quarter that at unit distance; at three times the distance, the intensity is one ninth, and so on. See Figure 3. When radiation encounters matter, one of three things can happen: It can be degraded to heat (absorption), it can change direction (refraction), or it can be reflected. See Figure 4.

The speed of light in a vacuum is 300 million meters per second. When light enters another medium—for instance, glass—its velocity changes. It slows down, but it resumes its former velocity on emerging from the denser medium. If it strikes the denser medium head on—perpendicular to the surface—the change in velocity has little effect, but if it strikes at an angle, the rays

Figure 3. Inverse square law. The intensity at *b* is 1/9 the intensity at *a*, because a given amount of light must cover nine times as much area at three times the distance. (Squares *a* and *b* are the same size.)

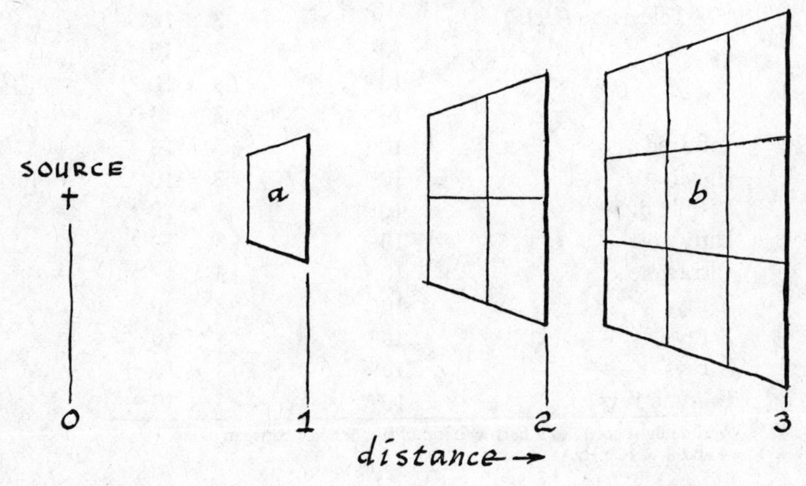

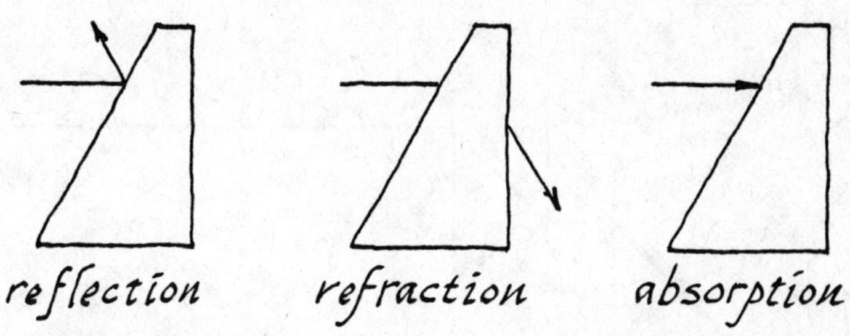

Figure 4. When radiation encounters matter, it may be reflected, refracted, or absorbed.

on one side are slowed before the rays on the other. The effect is like having one wheel in the mud—there is a change in direction. If the surface of the denser medium is curved, the change in direction will be different for every part of the surface. Lenses are based on this principle.

Chemical Composition. Matter is composed of molecules, which in turn are composed of atoms, of which ninety-two different varieties are found in nature. (A few others have been created in the laboratory.)

□ THE STATES OF MATTER

Matter can exist as a solid, a liquid, or a gas, depending on temperature, pressure, and chemical composition. The following statements constitute definitions of the three states of matter.

> The shape of a solid object is independent of its container. Matter in either of the fluid states (liquid or gas) will take the shape of its container.
>
> The volume of a liquid is independent of its container.
>
> A gas will assume the volume as well as the shape of its container.

In other words, gases are compressible, whereas liquids are not. Figure 5 is a plot of heat versus temperature for water. Look at

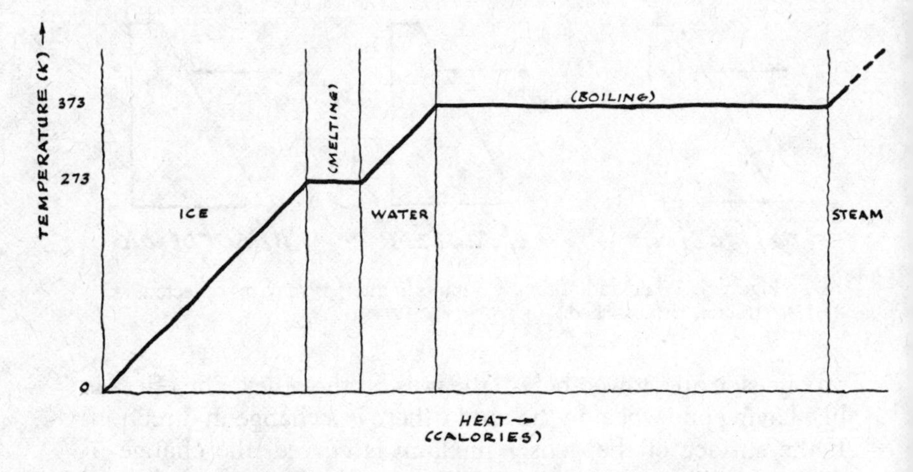

Figure 5. Plot of heat versus temperature for water. Note the discontinuities at 273 K and 373 K, corresponding to the freezing and boiling points, respectively.

the figure. You will notice that the plot is not a smooth curve but has two discontinuities, one at 273 kelvin and one at 373 kelvin, corresponding to the freezing and boiling points of water, respectively. The significance of this plot is this: As you apply heat to a volume of ice, water, or steam, the temperature rises in proportion to the heat applied, except during the transitions from solid to liquid and from liquid to gas indicated by the horizontal portions of the plot. Early investigators were puzzled to know what became of the considerable heat that is absorbed during these transitions. We now know that the heat is used in doing the work of changing the water from a solid to a liquid or from a liquid to a gas. That heat is returned to the environment when the steam condenses to water or when water freezes to ice.

Other substances behave similarly, but at different temperatures. Some substances (for instance, carbon dioxide) have no liquid state at ordinary pressures but go directly from a solid to a gas.

The freezing point is independent of pressure, but the boiling point of a liquid will vary, depending on pressure. The value of 373 kelvin (100 degrees Celsius) for the boiling point of water is taken at atmospheric pressure (100 kilopascals at sea level). Water boils at room temperature in a vacuum.

□ HYDRAULICS

Liquids are not compressible. Pressure applied to any part of a hydraulic system is transmitted equally in all directions. This is the way that hydraulic brakes work. The pressure of your foot is applied to the fluid in the master cylinder and is transmitted equally to all four wheel cylinders, thereby eliminating the dangerous swerving characteristic of mechanical brakes if not kept in perpetual adjustment. As mentioned earlier, gases, including air, are compressible. For this reason, hydraulic systems must include some provision for bleeding any air out of the system if they are to function properly.

□ PNEUMATICS

Crustacea creep about the ocean's bottom, oblivious to the enormous weight of the water above their heads. One reason that they don't think much about the water is that, because of the hydrostatic principles expounded in the preceding paragraph, the pressure is exerted equally in all directions; there is no net force influencing them to move in one direction or another. Similarly, we move about the bottom of an ocean of atmosphere, oblivious to the weight of the air and for much the same reason. But air has weight (being matter) and exerts at sea level a pressure of about fifteen pounds per square inch. This is fifteen pounds pressing on each and every square inch of our bodies, up, down, and sideways.

Gauges measure the difference between the pressure in a container and the pressure of the atmosphere; thus, when your tire gauge registers thirty pounds, the absolute pressure in your tires is forty-five pounds per square inch, partially offset by the fifteen pounds per square inch on the outside. Pressures less than atmospheric are called vacuums. (There is no such thing as the absence of all air, although interstellar space comes close—about one molecule per cubic meter.)

□ HEAT AND REFRIGERATION

Heating is simple. Our species has been at it for thousands of years. The technique is to burn something in the vicinity of where you want the heat, or use an electric heating element. If you are concerned with efficiency (or economy), you can limit losses by insulating the system from the environment.

Refrigeration is another matter. Here, you have the environment working against you; and insulation is as necessary to refrigeration as a boat is to bailing. (You know, if you were in the ocean without a boat, you'd have to bail like mad to stay afloat.)

To understand refrigeration, let's have another look at Figure 5. The problem is to get rid of heat. How about using ice? Not bad for a first attempt. Ice absorbs quite a bit of heat in going from the solid to the liquid state, and that heat can come only from its environment—meaning the immediate vicinity if the ice is inside an insulated container. When we throw out the melt water, we also throw away a lot of heat. This can be considered a manual refrigeration system.

There is another place on the plot of Figure 5 where there is a substantial change in heat without a corresponding change in temperature. This is in the transition from the gaseous to the liquid state. Recall that this transition is not a function of temperature alone but is also dependent upon pressure. Suppose there were a liquid that boiled at some convenient temperature below zero degrees Celsius at standard pressure and could be recondensed at room temperature, using moderate pressure. Several such substances do indeed exist. A synthetic substance, Freon, is used extensively as a commercial refrigerant.

Figure 6 is a schematic diagram of a refrigerator. The refrigerant gas is compressed by the compressor (a) and condenses in the condenser (b). On condensing, the refrigerant gives up considerable heat to the environment. Continued operation of the compressor forces the liquefied refrigerant past a restriction (c)

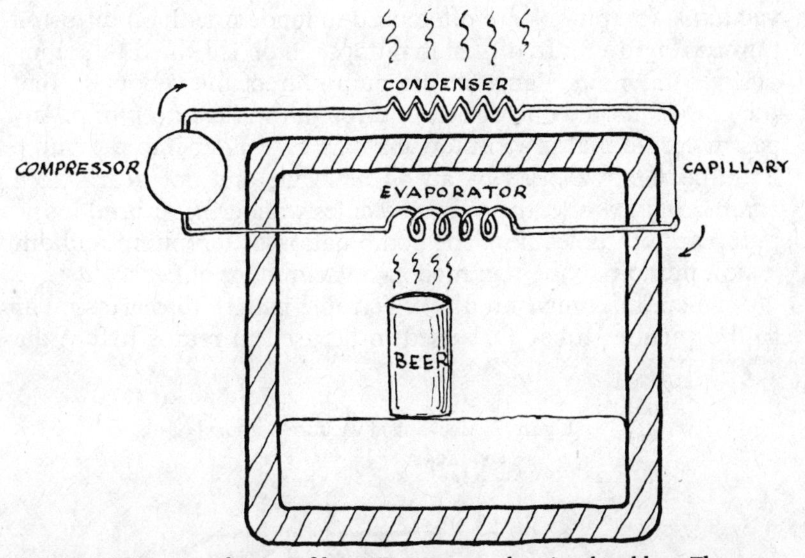

Figure 6. How to cool a can of beer. First you need an insulated box. Then you need a liquid such as Freon that boils at a very low temperature and can be liquefied by compression at room temperature, even on a very hot day. As the liquid boils, it absorbs heat from the interior of the box. The vapor is then compressed until its temperature is greater than that of the room. The vapor then condenses, giving up its heat to the room, and the cycle starts over again. The whole thing is a heat pump, pumping heat from the interior of the refrigerator out into the room. The capillary is there just to impede the flow of the liquid refrigerant so that the vapor can be compressed.

into the evaporator (d). The refrigerant boils in the evaporator and in so doing extracts considerable heat from the environment, which is the interior of the refrigerator. Then the cycle begins again: compression, condensation, and evaporation. The whole thing is a heat pump that pumps heat from the interior of the refrigerator out into the room.

□ VACUUM TECHNOLOGY

Rough vacuums are produced by mechanical vacuum pumps called roughing pumps or forepumps. The operating principle of mechanical vacuum pumps is illustrated in Figure 7. For higher vacuums, forepumps are often used in tandem with oil diffusion pumps (Figure 8). In diffusion pumps, an oil is heated to boiling and recondensed. Vanes in the pump direct the vapors so that they move in only one direction through the vacuum pump. Any gas molecule that is so unfortunate as to wander into the pump is immediately zapped by an oil molecule and not heard from again. After a while, the only molecules wandering around in the pump are oil molecules. The combination of forepump and diffusion pump may be regarded as a vacuum amplifier.

Vacuum is measured by electronic gauges that are similar to the vacuum tubes that used to be used in radios before the

Figure 7. Mechanical vacuum pump.

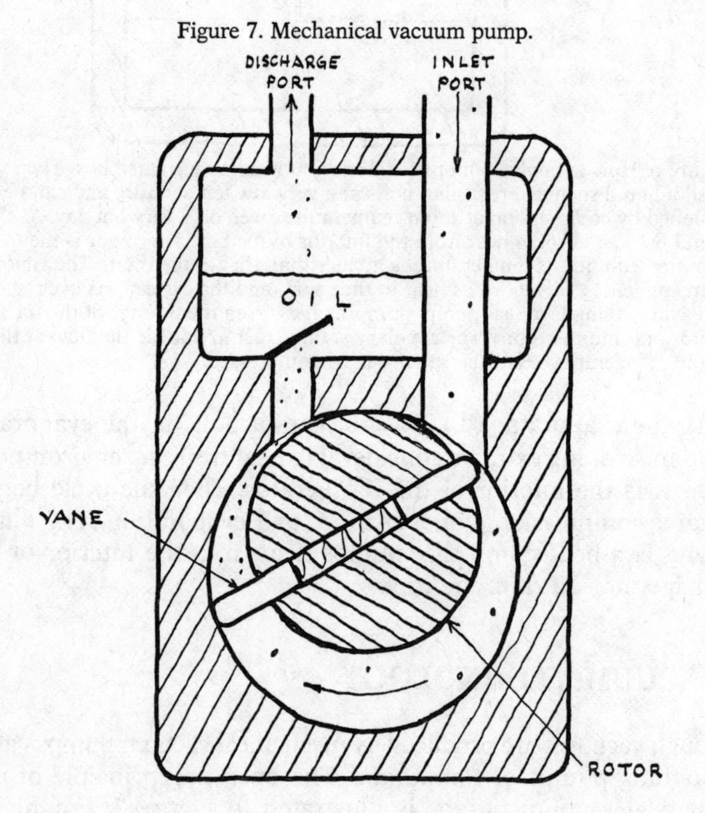

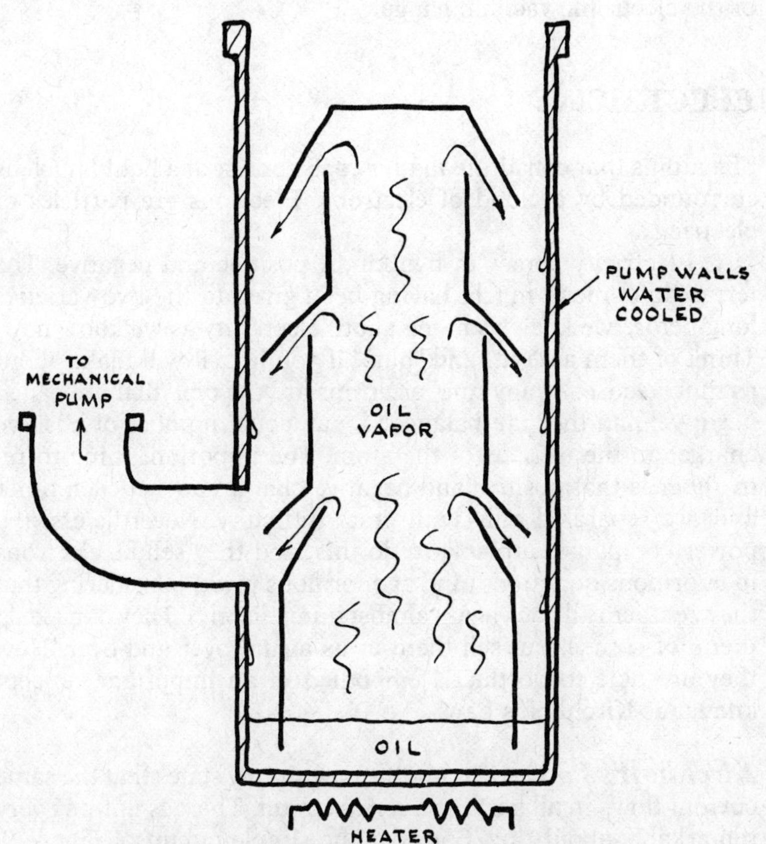

Figure 8. Diffusion pump. Oil in the bottom of the pump is heated to boiling. The rising vapors condense on the walls of the pump, which are water cooled, and the condensed oil runs back down to the bottom. The interior of the pump is evacuated by a mechanical vacuum pump (Figure 7). Stray molecules from the high-vacuum area that wander into the diffusion pump are entrapped by oil molecules and carried to the oil reservoir in the bottom of the pump and eventually evacuated by the mechanical pump.

advent of the transistor. A high-quality vacuum is essential to the proper operation of a vacuum tube. The effect of poor vacuum is measurable, and it is this effect that is the basis for operation of the electronic vacuum gauge.

□ ELECTRICITY

The atoms that constitute matter each consist of a liquid nucleus surrounded by a cloud of electrons. Electrons are particles of electricity.

Electricity comes in two kinds, positive and negative. The terms don't mean much, having been given to the two varieties long before we knew as much about electricity as we know now. Think of them as "red" and "blue" if you like; it will make about as much sense. At any rate, electrons are the type that are called negative; and they are balanced by an equal number of positive charges in the nucleus of the atom. The important thing to remember is that positive and negative charges attract each other and are separated only with great difficulty. Nevertheless, the power companies are able to do this, and they sell us electrons in enormous quantities (and at enormous price, considering that they get them all back again almost immediately). They don't keep them, of course, but sell them to us again, over and over. How they are able to do this is embodied in an important concept known as Kirchhoff's Law.

Kirchhoff's Law. Kirchhoff's Law simply states that the same current flows in all parts of a series circuit. There is nothing very remarkable about this. Consider the simple circuit of Figure 9, consisting of a battery, a switch, and a light. When the switch is open, no current flows. When the switch is closed, current flows from the battery through the lamp and the switch and back to the battery; there is no place else for it to go. This is called a series circuit. If the current through the lamp is one ampere, the current through the switch is one ampere, and so is the current through the battery.

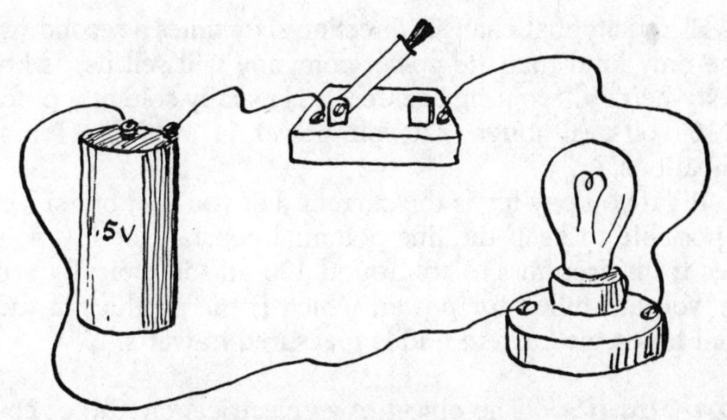

Figure 9. A simple series circuit.

Ohm's Law. Why is the current through the lamp one ampere, rather than 100 amperes, or half an ampere? Current and voltage are related by a law known as Ohm's Law. The light bulb resists the flow of electric current. This property is known as resistance and is measured in ohms. The battery provides an electromotive force called potential, which is measured in volts. Think of it as analogous to water pressure. Ohm's Law states that a potential of one volt will force a current of one ampere through a resistance of one ohm. Stated another way, a current of one ampere flowing through a resistance of one ohm will produce a potential drop of one volt.

If we have a twelve-volt battery and lamp resistance of twelve ohms, one ampere will flow through the circuit. If the lamp resistance is six ohms, two amperes will flow. This is a simple relationship.

Batteries produce a simple-minded kind of current called direct current. The power company sells us a complicated variety called alternating current. This simply means that the current changes direction periodically; but since we don't get to keep the electrons anyway, what does it matter? The light will burn just as brightly regardless of which direction the electrons are going, even if they change direction sixty times a second.

Electricity that changes direction sixty times a second (which is the only kind that the power company will sell us) is known as sixty-hertz alternating current. It is usually sold at a potential of 110 volts, although 220-volt power is available for some applications.

What you pay for is the current that you use; but since it is not possible to keep the line potential constant at 110 volts (it varies from around 110 to around 120 and is typically around 117), you are billed for power, which is the product of the potential times the current and is measured in watts.

Short Circuits. The object of an electrical circuit, of course, is to perform some sort of useful work. A conductive path that bypasses any part of a circuit is called a short circuit. Short circuits are not the only circuit defects; equally important are places where circuit continuity is lacking. Such defects are properly called open circuits.

Parallel Circuits. The circuits that we have been discussing are called series circuits, meaning that the electrons have to pass through each element of the circuit, one after the other, in order to get back to the source. The receptacles (outlets) in our houses are not wired like that, of course. If they were, we would have to have all of our appliances plugged in and turned on in order to have any one of them operating. Receptacles are wired across the power line in parallel (Figure 10).

Rating. The parallel circuits in our homes and offices are rated in terms of voltage and current. The voltage rating is the nominal (e.g., "110 volt") potential supplied at the outlet. The current rat-

Figure 10. Parallel circuits.

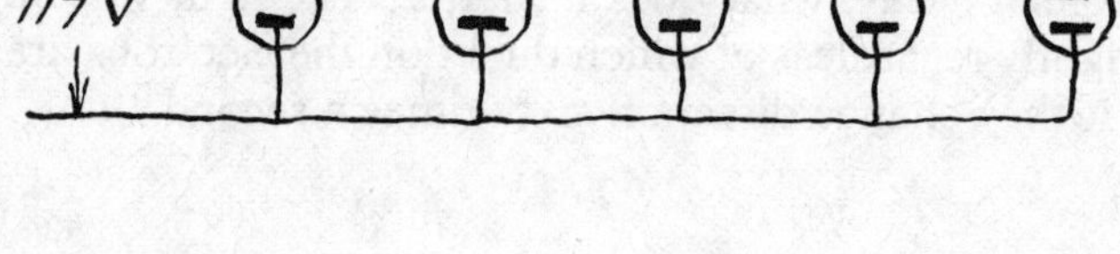

117 V

ing is the rating of the fuse or circuit breaker that is installed to protect the wiring from the dire effects of short circuits or overload. This, in turn, reflects the diameter of the wire that is used (smaller wires require smaller fuse values for adequate protection). Wires that are carrying too much current for their diameter can become hot enough to start fires.

Measurement. Potential is measured by placing a voltmeter in parallel with the circuit to be measured. Current is measured by placing an ammeter in series with the circuit to be measured. Resistance is measured by placing a battery and an ammeter in series with the resistance to be measured. These three functions are combined in the popular volt-ohm-milliammeter (VOM) or multitester, which combines one meter movement with a switch and various precision resistors to provide all three functions in one portable instrument. Another popular instrument for measuring potential is the digital voltmeter (DVM), which has a digital readout.

Induction. When an electrical current flows through a conductor, it generates a magnetic field. Nearby conductors are unaffected by the magnetic field as long as it is constant, but if there is a change in the field, a potential will be induced in the neighboring conductor. This phenomenon is known as induction. Alternating currents, of course, are continuously changing; and a sustained alternating potential will be induced in any nearby conductor.

Transformers are electrical devices designed to take advantage of induction for the purpose of coupling one circuit to another. To maximize the efficiency of the induction, the conductors are wound into coils around an iron core. The coils enable lots of conductors to be concentrated in a small volume, and the iron core serves to concentrate the magnetic field, since iron is more permeable to magnetism than are most substances. The coupling efficiency of a properly designed transformer can be very high.

The ratio of the potential in the secondary winding of a transformer to that in the primary winding will be equal to the ratio of the turns in the two windings. Thus, if there are 100 turns in the primary winding and 200 turns in the secondary, the secondary potential will be twice that of the primary. This would be a way, for instance, of deriving 220 volts from a 110-volt circuit.

The current drawn from the power line will be in inverse proportion to the turns ratio. Thus, in the 2:1 step-up transformer just described, if the current drawn by the load is one ampere, the current drawn from the line is two amperes (plus a small and usually negligible leakage current). The power in either circuit is volts times amperes, or 220 watts.

Rectification. Frequently, what is wanted in electrical equipment is not alternating current but direct current. Direct current is produced from alternating current by a process known as rectification, which makes use of two-terminal devices called diodes, which pass electrical current in one direction only.

Half Wave. A single diode will pass alternate half cycles of an alternating current. Such a device is known as a half-wave rectifier.

Full Wave. Two diodes can be arranged to pass both halves of the alternating-current waveform. Such a device is called a full-wave rectifier.

Bridge Rectifier. A very efficient full-wave rectifier uses four diodes in a bridge arrangement (named for its resemblance to the so-called "Wheatstone bridge" circuit for measuring resistance).

Filtering. Direct current obtained from a rectifier is not smooth like direct current from a battery but follows the contour of the alternating current half cycles. Filter capacitors placed across the direct current circuit charge up to the peak potential and discharge through the load between peaks, resulting in a much smoother direct current.

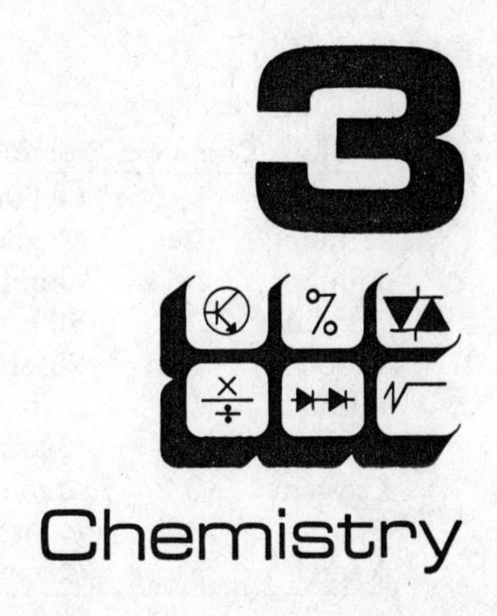

Chemistry

As we view the world around us, we are struck by the fact that there are many different kinds of substances, and mixtures of substances. There are, in fact, an enormous number of pure substances, called chemical compounds; but they are composed of a relatively few different kinds of atoms.

A chemical compound is a substance that is composed of various atoms in very definite proportions. For instance, water consists of hydrogen and oxygen in the exact ratio of two atoms of hydrogen for every atom of oxygen. The chemical formula for this compound is H_2O, where the H_2 stands for the two atoms of hydrogen and the O stands for the atom of oxygen.

There are ninety-two different kinds of atoms found in nature, some in extremely small quantities, plus a few that have been created in the laboratory. Since there are more than 100 chemical elements and only twenty-six letters, most of the symbols for the chemical elements consist of two letters—a capital letter and a lower-case letter. The first thirty elements of the periodic table, comprising 99.9995 percent of all the matter in the universe, are listed with their symbols in Table 2.

Hydrogen and helium, the first two, comprise more than 99 percent of all the atoms in the universe but only 1.2 percent of the atoms in the earth. The first twenty-six (down to iron) con-

Table 2. Chemical Elements and Symbols.

Hydrogen	H	Sodium	Na	Scandium	Sc
Helium	He	Magnesium	Mg	Titanium	Ti
Lithium	Li	Aluminum	Al	Vanadium	V
Beryllium	Be	Silicon	Si	Chromium	Cr
Boron	B	Phosphorus	P	Manganese	Mn
Carbon	C	Sulfur	S	Iron	Fe
Nitrogen	N	Chlorine	Cl	Cobalt	Co
Oxygen	O	Argon	Ar	Nickel	Ni
Fluorine	F	Potassium	K	Copper	Cu
Neon	Ne	Calcium	Ca	Zinc	Zn

stitute 99.99 percent of the earth's crust. Finally, hydrogen, carbon, nitrogen, and oxygen comprise 99 percent of the atoms in the human body.

The elements are classed as metals or nonmetals according to whether or not they conduct electricity. Most of the elements are metals, but the nonmetals are of equal importance because of their abundance in nature. Some typical metallic elements are iron (Fe), copper (Cu), gold (Au), silver (Ag), and lead (Pb). Some typical nonmetallic elements are carbon (C), sulfur (S), phosphorus (P), oxygen (O), and nitrogen (N).

Carbon is unique among the elements in being able to form complex compounds with itself. For this reason, the number of carbon compounds far exceeds the number of all the noncarbon compounds put together, and the science of chemistry is divided into two branches called organic chemistry, which is the chemistry of carbon compounds, and inorganic chemistry, which is all the rest. Inorganic chemistry, being simpler, will be discussed first.

□ INORGANIC CHEMISTRY

The chemistry of all the elements except carbon is called inorganic chemistry.

Gases. Some of the simplest compounds in nature are gases at ordinary temperature and pressure. Several of the nonmetallic elements exist in the form of diatomic molecules—for instance, oxygen (O_2), nitrogen (N_2), and hydrogen (H_2). (The subscript 2 indicates the number of atoms in the molecule.)

For reasons that will not be discussed here, some elements are so inert that they will not form compounds, even with themselves. These are the so-called "inert" or "rare" gases: helium (He), neon (Ne), argon (Ar), xenon (Xe), and so on.

Some simple gaseous compounds are carbon dioxide (CO_2), ammonia (NH_3), and methane (CH_4). CO_2 indicates one carbon and two oxygen atoms. NH_3 indicates one nitrogen and three hydrogens. CH_4 indicates one carbon and four hydrogens.

Solutions. Many compounds dissociate in water solution to form ions—charged particles—which means that an electron that belongs to one atom has gone off with another. This sort of infidelity is very common in chemistry. Recall that an electron is a unit of negative electricity. Hydrogen—the simplest element—is composed of one proton and one electron. The proton has all the mass and a positive charge; the electron has very little mass and a negative charge. Hydrogen atoms are always coming apart in solution to form protons (H^+) and electrons (e^-). The protons move around in the water, but the electrons usually remain attached to any molecular fragment that has lost a proton.

Metals in general tend to lose an electron in solution and form positive ions. Nonmetals tend to gain an electron and form negative ions. Compounds that tend to lose a proton in solution are called acids. Compounds that tend to grab protons are called bases.

☐ ORGANIC CHEMISTRY

Carbon atoms have a tendency to form chains (aliphatic compounds) and rings (aromatic compounds).

Carbon atoms have four hooks or bonds by means of which they are able to grab onto other atoms. Methane (CH_4), which was mentioned previously under "Gases," is the simplest of the organic compounds; and its radical, CH_3-, is very common in organic chemistry. Indeed, it is so common that it is sometimes abbreviated Me, as if it were an element.

Next to the methyl radical (CH_3-), the six-membered ring (C_6H_5-) is the most common structure in organic chemistry. Heterocyclic rings (five- and six-membered rings containing atoms other than carbon atoms) are also fairly common, especially in biochemistry.

□ BIOCHEMISTRY

Living things, from viruses to trees, consist of complex chemical systems. The major categories of nutrients—fats, carbohydrates, proteins—are also the major categories of living matter.

Proteins are the most complex of all chemicals, but they are built up of simpler units called amino acids, of which there are about twenty, just as words are made of letters. But unlike words, even the simplest proteins contain hundreds of amino acids; and the more complex ones may contain millions. The twenty amino acids and their symbols are enumerated in the following list. Note that the symbols follow the general rule for atoms and other parts of molecules: the first letter is a capital, and the following letters are lower case.

Alanine	Ala	Leucine	Leu
Argenine	Arg	Lysine	Lys
Asparagine	Asn	Methionine	Met
Aspartic acid	Asp	Phenylalanine	Phe
Cysteine	Cys	Proline	Pro
Glutamine	Gln	Serine	Ser
Glutamic acid	Glu	Threonine	Thr
Glycine	Gly	Tryptophan	Trp
Histidine	His	Tyrosine	Tyr
Isoleucine	Ile	Valine	Val

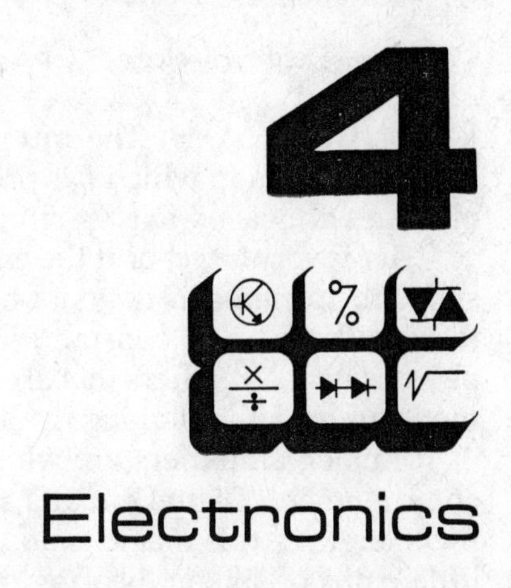

4 Electronics

A great deal of technical writing has to do with electronic circuits. If you write service manuals, you will need to know much more about electronics than this book can teach you. If you write operator manuals, you don't really have to know a whole lot more than the operator; and you just might get by with the light introduction to electronics presented in this chapter. To be on the safe side, read a good introductory textbook—or, better, take a course in electronics at your local community college.

Electricity is the term that we use when speaking of large currents, such as are available at the receptacle in the wall. When speaking of small currents—milliamperes and microamperes rather than amperes—the appropriate term is *electronics*.

□ AMPLIFICATION

The central idea of electronics is amplification—the use of a small current to control a larger one. Amplification can be used to control the amplitude of an alternating current (analog electronics) or to switch a direct current on and off (digital electronics).

Consider the three-terminal device depicted (very schematically) in Figure 11. A current *i* flows from *A* to *B*. Another current

I flows from C to B. The ratio I/i is known as the gain of the device. Circuits in which I is proportional to i over a wide range of values are called analog circuits.

If i is very large, or if the gain is very large, the power source supplying terminal C may not be able to provide enough current to keep the ratio I/i constant, in which case the input is said to be saturated. Amplifiers that are designed to work in the saturated mode are called switching or digital amplifiers.

Analog amplifiers are widely used in communications and entertainment. Digital circuits are widely used in computers and the like. All of this magic is done with integrated circuits (ICs)—little chips of silicon, bearing microscopic circuits—each equivalent to thousands of transistors. Each chip is enclosed in a plastic case with leads so that it can be soldered into larger assemblies called circuit boards.

Signal levels in digital circuits commonly assume one of two possible values, variously called "zero" and "one," "true" and "false," or "high" and "low." These terms have about as much meaning as "positive" and "negative," which is to say, none. We shall use the terms *high* and *low*.

□ INVERSION

Consider the simple circuit of Figure 12, consisting of a battery, B, a resistor, R, and a switch, S. When switch S is open, the potential at point C is the same as that at the positive terminal of the battery—five volts. (Since no current is flowing, there is no voltage drop through the resistor.) A voltmeter connected across points C and B registers five volts. When switch S is closed, point

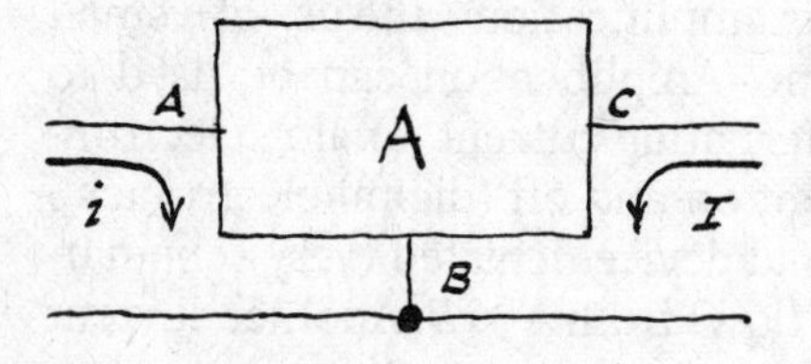

Figure 11. Amplification. A small current through terminal A controls a larger current through terminal C.

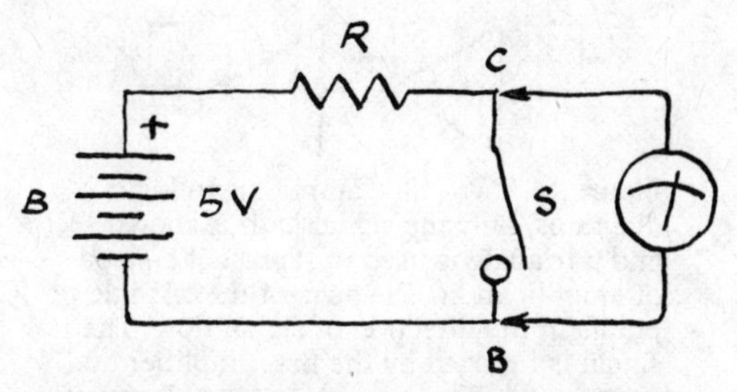

Figure 12. Simple circuit consisting of a battery, B, a resistor, R, and a switch, S. When the switch is open, the meter reads five volts. When the switch is closed, the meter reads zero volts.

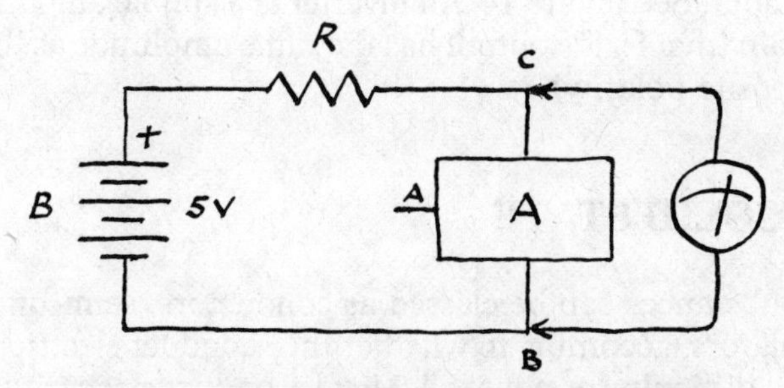

Figure 13. Simple amplifier circuit. The amplifier A replaces the switch in Figure 12. When there is no signal at terminal A, the meter reads five volts. When there is a positive signal at terminal A, the meter reads zero volts.

C is physically connected to point *B*, and the voltmeter registers zero volts.

Now replace switch *S* with an amplifier (Figure 13). Current flowing through terminal *A* has the same effect as closing the switch in Figure 12: the voltage at point *C* decreases to zero. If current flows through terminal *A* for one second, then the voltage at terminal *C* will be zero for one second. This inversion of the signal is characteristic of single-stage amplifiers, both analog and digital. If two amplifiers are connected in tandem, a second in-

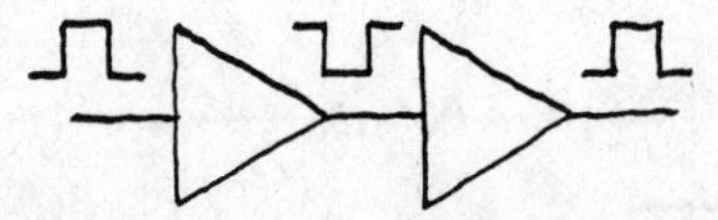

Figure 14. Two-stage amplifier. In logic diagrams, only the signal path is shown, and a triangle is used to represent a stage of amplification. The apex of the triangle points in the direction of signal flow. The signal is inverted by the first amplifier and reinverted by the second, so the polarity of the output is the same as that of the input.

version takes place, and the resulting signal has the same polarity as the input. See Figure 14. An inverter is a simple amplifier with unity gain (that is, the output has the same amplitude as the input but opposite polarity).

□ THE SOLID STATE

Solid substances can be classed as conductors, semiconductors, or insulators, according to whether they conduct electricity readily, with difficulty, or not at all. Metals conduct electricity readily, and the conducting pathways in electrical and electronic circuits are made of metal. Most other substances (plastics, glass, ceramics) are nonconductors of electricity, or insulators. A few materials (diamond, germanium, silicon) are classed as semiconductors. In semiconductors, a small current can be made to control a larger one. It is this phenomenon that makes modern electronics possible. In this chapter, we shall discuss some of the aspects of electronic equipment that are most readily visible to the user.

□ POWER SUPPLIES

The alternating current available at the wall receptacle is not suitable for operating electronic circuits. What is required is fairly high-quality direct current, frequently at several different voltages. A fuse or circuit breaker is used to protect the power supply wiring, just as fuses and circuit breakers are used to protect the wiring of a building. A transformer is used to isolate the equipment from the power line and to transform the voltage to a potential that is suited to the circuitry that is to be powered. The secondary voltage from the transformer is rectified and filtered and passed through a series pass element called a regulator, which is an amplifier that regulates the voltage within a specified tolerance.

□ CONNECTORS

Seldom is a piece of electronic equipment complete unto itself like a pocket calculator; usually it is used in conjunction with other devices. Connectors provide a means of connecting (and disconnecting) cables that go to other equipment. Multipin connectors are used with multiconductor cables. Coaxial connectors are used with coaxial cables—cables in which a center conductor is concentric with an external shield.

□ CONTROLS AND INDICATORS

Controls and indicators can be divided into two classes: analog and digital.

Controls. Manual controls permit human interaction with an electronic circuit. Two kinds of controls are commonly used.

Analog or continuously variable controls are usually variable resistors called potentiometers. (Colloquially, they are known as "pots." A resistor is a circuit element that resists the flow of alternating and direct currents.) Potentiometers are used to adjust the value of an electrical potential or current.

Digital controls (switches) have a finite number of mechanical positions. Rotary (selector) switches may have as many as twelve positions. Toggle and rocker switches are usually binary (two-position) switches, although some may have a center "off" position; the wall switches in your home are binary switches. Push-button switches are also binary switches. Alternate-action push buttons change state each time they are pressed. Momentary push buttons are active only while the button is held in the depressed position. Sometimes several push buttons are ganged mechanically so that only one can be activated at a time; pressing one releases any other button that is depressed. In this way, ganged push buttons can duplicate the function of a rotary selector switch. Illuminated push buttons contain a light that indicates the switch condition.

Displays. Displays are used to convey to a human operator information concerning the status of electronic equipment.

Meters are analog devices that have a moving pointer which can be used to indicate the value of some voltage or current (or of some parameter that can be converted to a voltage or current). Light-emitting diodes (LEDs) are digital devices that are used to indicate the presence or absence of some condition. LEDs can be arranged in patterns to simulate numbers or letters, thereby increasing the amount of information that can be conveyed.

Cathode ray tubes (CRTs) use an electron beam to produce moving images on a phosphorescent screen, thereby conveying much more information than can be conveyed by any pattern of LEDs. The television picture tube is a familiar example of a CRT.

□ DIAGRAMS

It is quite impossible to comprehend complex electronics circuits from textual descriptions alone; circuit diagrams are absolutely essential. Several kinds of diagrams are commonly used.

Block Diagrams. Block diagrams are the simplest and frequently all that are needed to give the reader a superficial understanding of how the equipment works. In block diagrams, labeled rectangles are used to indicate major functional areas of the circuit. Only the signal paths are indicated; the power buses are omitted for clarity (unless, of course, the block diagram is a diagram of a power supply).

Timing Diagrams. Timing diagrams are sometimes used with digital circuits to help the reader appreciate what is happening in one part of the circuit simultaneously with an event in another part. Timing diagrams are really charts in which the horizontal axis is a time line that is common to the waveforms that constitute the body of the diagram.

Logic Diagrams. Logic diagrams (used only with digital circuits) are the next step in complexity after block diagrams. Logic diagrams use special symbols to indicate the logical functions AND, OR, NAND, and NOR. A triangle with the apex pointed in the direction of signal flow is used to indicate amplification of the signal. A small circle attached to a symbol indicates inversion of the signal. Integrated circuits (ICs) with special functions are sometimes indicated as rectangles with multiple inputs and outputs. Logic diagrams frequently contain all the information that is necessary for troubleshooting digital circuits at the circuit-board level. Gate circuits perform the logical functions of AND, OR, and NOR. Each gate has two or more inputs and a single output.

AND Gates. An AND gate (Figure 15) indicates that a signal must be present at both inputs for a signal to be present at the output.

OR Gates. An OR gate (Figure 16) indicates that a signal present at either input is sufficient for a signal to be present at the output.

NAND Gates. A NAND gate (Figure 17) indicates that a signal must be absent at both inputs for a signal to be present at the output.

NOR Gates. A NOR gate (Figure 18) indicates that the presence of a signal at either input is a sufficient condition for the absence of a signal at the output.

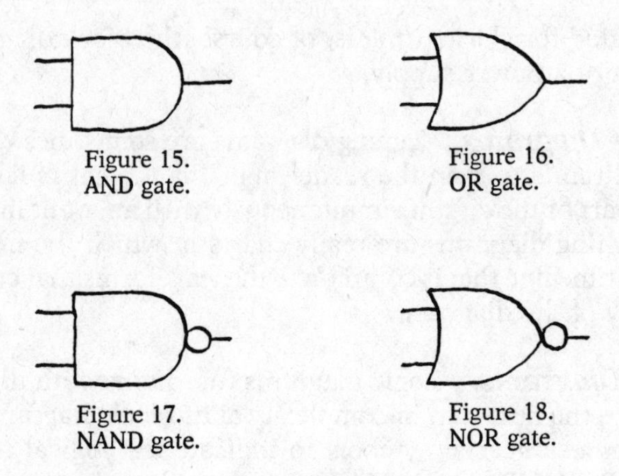

Figure 15.
AND gate.

Figure 16.
OR gate.

Figure 17.
NAND gate.

Figure 18.
NOR gate.

Flip-flops. A flip-flop is a circuit that changes state with each input pulse. It works like a ballpoint pen—press once and the point comes out; press again and the point retracts.

If two amplifiers are connected in such a way that the output of one is connected to the input of the other and vice versa, the result is a device that has two stable states. See Figure 19. If input *A* is high, output *B* will be low, input *C* will be low, and output *D* will be high, reinforcing input *A*. The device will remain in this state until reset by a low pulse at input *A* or a high pulse at input *C*.

Counters. A bunch of flip-flops connected together is a counter. There is no output from the second flip-flop until the first flip-

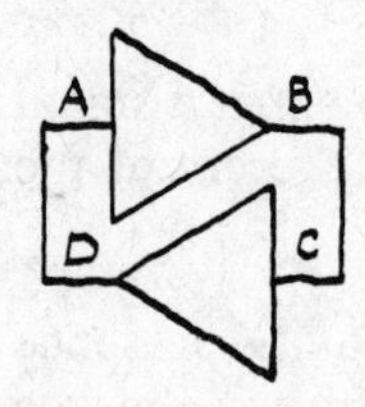

Figure 19. Flip-flop. If input A is high, output B will be low, input C will be low, and output D will be high, reinforcing input A. This device has two stable states. A high pulse applied to a low input or vice versa will cause the device to flip (or flop) to the alternate state.

flop has been pulsed twice, and so on. Three flip-flops give an output on the eighth pulse, and so on.

Clocks. Proper functioning of digital circuits depends on precise timing. A clock circuit generates a regular timing pulse, and gates are connected so as to ignore pulses that do not arrive during specified time intervals.

Schematic Diagrams. Analog circuits require a schematic diagram for total comprehension. Discrete components (resistors, diodes, capacitors, transistors, and so on) as well as integrated circuits are shown on schematic diagrams. Electronic components are used to control electric currents, direct them hither and yon, and generally make them do what we want them to do.

Resistors. Ohm's Law, as you may recall, defines the relationship of current, potential, and resistance in an electrical circuit. When it is desired to limit the current or drop the potential in a circuit, a device called a resistor is used. Resistors are fairly common in electronic circuits and are relatively cheap. They are usually made of carbon. Carbon, although it is a conductor, is not an especially good conductor. Resistors vary in value from a few ohms to several megohms. Resistors work by converting electrical power to heat. If considerable power is to be dissipated, it is important to specify the power rating of the resistor (in watts) as well as its resistance value. Resistors with high power ratings are physically larger than those with lower power ratings.

Capacitors. Capacitors serve to separate alternating currents from direct currents. A capacitor consists of a piece of insulating material sandwiched between two conductors. As far as direct currents are concerned, a capacitor is an open circuit. In an alternating-current (ac) circuit, electrons piling up on one of the conductors during half of the cycle repel an equal number of electrons from the other conductor. The situation is reversed during the alternate half cycles. In this way, capacitors block direct currents while effectively passing alternating currents. An important use of capacitors in electronics is for shunting unwanted ac signals ("noise") to ground. (*Ground* is the zero-potential reference level in electronic circuits.)

Inductors. Inductors perform a function that is the opposite of the function of a capacitor; that is, they pass direct currents while impeding the flow of alternating currents. An inductor consists of a conductor wound into a coil. As electrons traverse each turn of the coil, they induce an opposite current in the next turn. Inductors are comparatively expensive, so engineers plan their circuits so as to depend principally on capacitors for separating direct from alternating currents.

Diodes. Diodes are semiconductor devices that will pass current in one direction only; hence, they are useful for separating positive from negative pulses and steering direct currents where you want them to go. Alternating currents will emerge from a diode as a series of positive or negative pulses, depending on which way the diode is installed.

Transistors. The devices that started the solid-state revolution are still used in places where it is necessary to control larger currents than can be borne by integrated circuits. A transistor is a three-terminal device in which a small current through one terminal is used to control a larger current through another terminal. (The third terminal is common to both the input and the output circuit.) Depending on the circuit configuration, a transistor can act as either a switch or an amplifier.

Relays. A relay is a switch operated by an electromagnet. It is used to control larger currents than can economically be controlled by transistors. Relays are usually driven by transistors.

Integrated Circuits. Integrated circuits (ICs) are the workhorses of contemporary electronics. The circuit equivalent of thousands of transistors can be integrated in one IC. IC operational amplifiers require the addition of a few components to perform a variety of tasks requiring high-gain direct-current (dc) amplification.

Wiring Diagrams. Wiring diagrams are partially pictorial diagrams showing the ends of the wires that are used in a piece of electronic or electrical equipment. The destination of each wire is indicated in an abbreviated form. Wiring diagrams are indispensable for advanced troubleshooting and repairs.

Circuit Board Layouts. A circuit-board layout diagram is a pictorial diagram of a circuit board showing all of the components.

5 Computers

A computer is a remarkable machine—part hardware and part moonshine. Without the software that makes it go, the hardware just sits there, unblinking and uncomprehending. And the software—what is that? Invisible and incomprehensible ones and zeroes in a magnetic medium, inert until matched with the proper hardware.

Yet the software is machinery as much as the hardware is. It is an extension or continuation of the computer's memory. It contains switches that can be turned on and off. If you have enough hardware to interface with the real world—a terminal, a central processing unit (CPU), a disk drive, and a printer—you can design and build machinery of unlimited complexity at virtually no cost except for your time. And the software—where is it? On the disk, surely, but also in the computer's memory.

□ HARDWARE

Just as a brain needs lungs and bones, eyes, ears, arms, and a heart in order to function, so does a program need a modicum of hardware. The following items are considered indispensable.

The Central Processing Unit. The incredibly fast central processing unit is the real computer; the rest of the hardware are peripherals. In essence, two things serve to distinguish the multimillion-dollar mainframe computer from the Apple of your I: the speed of the CPU and the size of its memory. In almost all other respects, one is as good as the other. If you are not content with microsecond response times but require picosecond response, you will have to pay for it. Similarly, if 128K of memory (about the limit of most personal computers) is inadequate, you can get more—at a price.

Memory is measured in binary digits or bits. A bit is a quantum of information; it is the smallest amount of information there is; it is the difference between yes and no. Paul Revere's lanterns represented two bits of information. When the first lantern was lit in the steeple of the Old North Church, one bit of information was broadcast. By prearrangement, the Minute Men then knew that the British were coming. They then had to wait for the second bit of information to know whether the British were coming by land or by sea. (Note that the absence of a signal also carries information.) This is the essence of digital communication. Bits are organized into multiples called bytes. A byte is eight bits.

A more convenient unit to use in describing digital memories is the K. One K equals 2^{10}, or 1024 bytes. (This is not to be confused with the metric prefix *k*, which means "times 1000.")

Computer memory comes in two styles: read-only memory (ROM) and random-access memory (RAM). Read-only memory (also called firmware) can be read by the CPU but cannot be modified. Read-only memory is handy for storing the basic information that the computer needs each time that it is turned on, like where to go for stuff and how to talk BASIC. Some ROM is programmable (outside the computer), and some programmable ROM can be erased and used over (EPROM). Read-only memory (ROM) is the brains of the computer. It is the part that knows what to do—and where to go for help—when the computer is first turned on.

The computer uses random-access memory as a scratch pad. Information may be stored in RAM and retrieved as required.

Random-access memory is like a blank book. You can open it at random to any page and write on (or read) any line. Memory size is an important measure of computer power. When we say that a computer has 32K of memory, we are generally speaking of RAM.

The Terminal. People and computers interact via a terminal. A terminal is actually two devices: a keyboard and a cathode ray tube (CRT). Although they may be housed in the same cabinet, there is no necessary connection between the two, except via the CPU. The person sends messages to the computer via a typewriter keyboard. The computer sends messages to the person via the CRT, which is similar to the picture tube in a television set.

When we type a message on the keyboard, it appears instantaneously on the screen; it is as if we are typing directly on the CRT, but each character is actually an echo from the CPU. For instance, if you type the letter *A*, the CPU reads that letter and sends its own *A* to the CRT. This happens so fast that you think that it is your *A* that you see on the screen, but it is actually an *A* generated by the CPU.

Bulk Storage. Vast as the CPU memory may be (in some computers), it is not infinite. In addition, it is volatile; when you turn the power off, the RAM forgets everything it ever knew. For these reasons, a computer has need of a permanent storage place for more information than can be contained in 32K or 48K of RAM.

Disks. Magnetic disks are the most common media for bulk storage. Large and medium-sized computers use rigid disks. Small computers use flexible "floppy" disks.

Magnetic disks are shaped like phonograph records, but they work like magnetic tape. The disks rotate at high speed, and the read/write heads, which are similar in design to the magnetic heads in a tape recorder, only smaller, move across the disk to record or pick off information. Except for those used with the flexible disks, the heads do not actually touch the disk but float on a cushion of air only a few microns thick. Any foreign particles, such as dust or even cigarette smoke, are sufficient to cause a "crash" and loss of the disk and possibly priceless data. For this reason, disk drives are pressurized so that any leaks will cause air to move out from the drive, rather than in. (An inward movement of air would carry the risk of contamination by dust and smoke.) The air that is used to pressurize the drive is, of course, filtered.

"Winchester" drives are sealed drives, serviceable only by the manufacturer. In Winchester drives, the heads are even closer to the disk, making possible an information packing density several times that of conventional disk drives. (The first Winchesters were designed to hold thirty megabytes of information on each side. The code name for the project was "30-30,"* which some wag nicknamed "Winchester." The name stuck.)

Tape. Mankind's perennial problem is what to do with stuff that's not worth keeping but too good to throw away. With paper, we use cardboard transfer files and put them in the attic. With computer data, we use a cheaper magnetic medium. We put our obsolete magnetic records on tape. Tape is slow and lacks random access, but it will hold an enormous amount of data at a cost of about 80 cents per megabyte.

Printers. The time will come when one will want to show one's data to someone else. Only a limited number of people can gather around the CRT. What is wanted is a "hard copy." Two kinds of printers are in common use.

Dot Matrix. Dot-matrix printers are very fast. The characters are formed by tiny dots. The print quality is good enough to show to one's friends and family.

*A rifle that fires a .30-caliber cartridge having a 30-grain powder charge; the Gun that Won the West.

Formed Character. Formed-character printers are rather slow (but still faster than typing). The printout is equivalent to that gotten from a good quality electric typewriter. Each character is fully formed, as it is with a typewriter element.

□ SOFTWARE

Software for computers comprises programs and data. Programs are the instructions that tell a computer what to do. Data are numbers.

Programs may be divided into two categories: those that you knew you needed before you bought the computer and those that you found out that you needed after you bought the computer. Programs of the first category are called applications. (Presumably, you didn't buy a computer because of the blinking lights; you bought it because you had a specific application—or specific applications—in mind.)

Programs of the second category are called system software. They make the system easier to use. In many instances, the system will not operate at all without a program called an operating system. The outstanding characteristic of system software is that you wouldn't need it if you didn't have a computer.

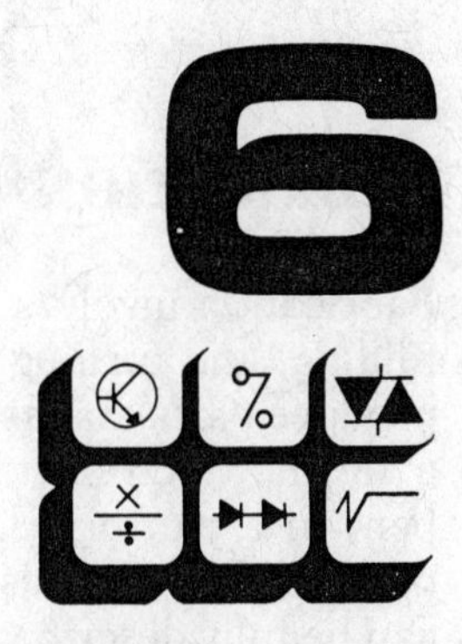

Industrial Processes

Metals and plastics are the principal raw materials used in light industry. These materials are generally available in the form of sheets and rods and extruded shapes, such as tubing. Metal castings and molded plastic parts are used for many high-volume applications. Castings require some machining before they can be used.

□ MECHANICAL DRAWING

Before a part can be fabricated, the engineering department must prepare a mechanical drawing that completely describes the part. Some parts, for instance wires and tubes, can be completely described by merely specifying the material and the length. The "drawing" in these instances consist entirely of words and numbers. Most parts require from one to three principal views; and some require auxiliary views, such as cross-sections, details, and the like. Flat parts, as well as parts in which two or more views would be identical, such as cylinders and spheres, require only one view for a complete description. Parts that are assembled from two or more pieces require assembly drawings.

□ FABRICATION

Fabrication involves cutting, bending, boring, drilling, grinding, milling, and turning operations performed on raw materials to produce parts for assemblies. Sheet metal parts are cut to size in a shear, which operates on the same principle as scissors, and bent in a pan brake. Special jigs may be required for bending parts made from rods or extrusions. Some three-dimensional parts can be fully described in one cross-sectional view. A lathe is used for making such parts.

Holes. Holes in sheet metal parts are made by punching. Cylindrical parts are bored in a lathe. A drill press is used for making holes in other three-dimensional parts.

Shaping. A rotating abrasive wheel is used for shaping some three-dimensional parts. A milling machine is used for most other shaping operations that cannot be accomplished by bending, boring, drilling, or grinding.

□ FINISHES

Metal parts, unless made of stainless steel, require some sort of finish to enhance their appearance and protect them from corrosion. Cabinets and panels are usually painted. Smaller parts may be plated with a corrosion-resistant metal such as cadmium, chromium, or zinc. Aluminum parts may be given a chemical treatment (anodizing or alodyning).

□ ASSEMBLY

Large steel structures, such as cabinets, are usually welded. Electric spot welding is used if the stress is expected to be minor. Brazing uses melted brass to hold parts together. Solder (a mixture of tin and lead) is used for electrical continuity where mechanical strength is not required.

Riveting. A rivet is a piece of malleable metal that is used to hold two pieces of material together. The rivet is placed through matching holes in the two parts and then deformed so that the two pieces are clamped together.

Threaded Fasteners. A bolt is a threaded fastener designed to be used with a nut and washer to hold two pieces of material together. A machine screw is distinguished from a bolt only by the nature of the use to which it is put. (If it is used with a nut, it is called a bolt; if it is used in a tapped hole, it is called a screw.) A screw that has a head is called a cap screw. A screw that has a socket for a tool is called a socket screw. A number of parameters are required in order to describe a screw.

Diameter. Screws smaller than 1/4 inch in diameter are described according to the American Wire Gauge (AWG) of their diameter. Typical screw diameters are number 2, number 4, number 6, number 8, and number 10. Number 6 screws are perhaps the most common. Screw diameters 1/4 inch and over are described in inches.

Pitch. The number of threads per inch of a screw is called the pitch. Number 2 screws usually have 56 threads per inch. Number 4 screws usually have 40 threads per inch. Numbers 6, 8, and 10 screws generally have 32 threads per inch. One-quarter-inch machine screws generally have 20 threads per inch. A number 6 screw with 32 threads per inch is called a "6-32 machine screw."

Length. The length of a screw is qenerally given in inches. A 6-32 screw one inch long is described as a "6-32 × 1 in." machine screw.

Head. Slotted-head screws require a flat-bladed screwdriver. Phillips-head screws require a Phillips screwdriver. Hex-head screws require a wrench. Hexagonal socket screws (colloquially called "allen screws") require a hex key or "allen wrench." Screw heads are further categorized according to their profiles. See Figure 20.

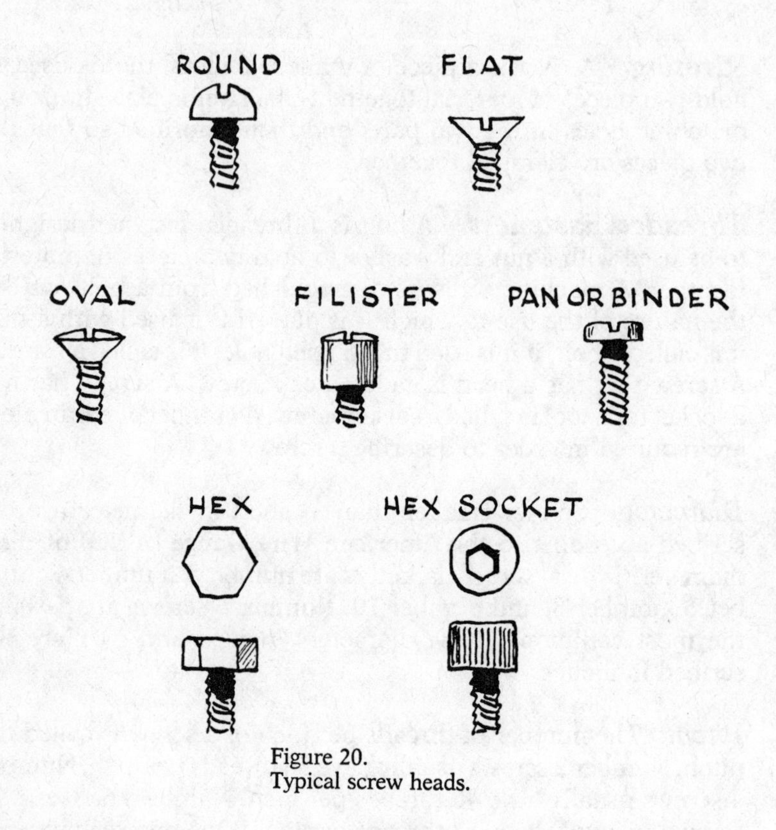

Figure 20.
Typical screw heads.

Material and Finish. Screws may be made of brass, steel, or nylon. Metal screws may be plated with zinc, cadmium, or chromium.

Hand Tools. Screwdrivers, wrenches, pliers, and soldering irons are used in assembly operations.

Screwdrivers. Flat-bladed and Phillips screwdrivers are used in assembly. An important measure of a flat-bladed screwdriver is the width of the blade. One-eighth- and 1/4-inch blades are commonly used.

Wrenches. Open-end wrenches are used for turning nuts and hex-headed cap screws. Adjustable end wrenches (colloquially called "Crescent" wrenches) will fit a large range of sizes. Box wrenches have twelve (or more) notches to fit the corners of the nuts and screw heads and are very handy where space is limited; but being closed, they have to fit over the end of the part that is to be turned and hence cannot be used with tubing. Socket wrenches are two-part tools consisting of a socket, which is like the end of a box wrench, and a detachable handle. One handle will fit an entire set of sockets. In addition, various handles are available for various situations. Hex keys ("allen wrenches") are used to turn hex socket screws. Pipe wrenches have teeth that can grip a pipe.

Pliers. Gas pliers are handy in an emergency when the proper tool is not at hand, but they are likely to mar the parts that they are used on. Diagonal cutting pliers ("dikes") are used for cutting wires. Long-nose pliers are used for handling small parts and bending wires.

Wiring and Plumbing. For major systems, most of the wiring is cabled together in a wiring harness, which is designed and installed as a subassembly, complete with connectors for circuit boards and external cables.

Tubing is used to route fluids to, through, and from equipment. Two methods are commonly used to connect tubing to other parts in a system. In one method, a nut and ferrule are slipped over the tubing, and the end of the tubing is then flared to mate with a conical surface on the mating part. The nut is then threaded onto the mating part and tightened in order to press the flared end of the tubing against the conical projection. Another method uses a soft ferrule that deforms under pressure to make a hermetic seal with the tubing.

Pressure regulators are used to regulate the pressure of gases. Gauges measure the pressure of fluids. Valves are used to control the flow of fluids.

Power Supplies. The electrical power that comes out of the wall is never the proper potential for electronic circuits. For this reason, some sort of power supply is always needed.

Line Cord. An important parameter concerning the line cord is its length. Another is the kind of connector on the end.

Fuses and Circuit Breakers. Fuses are used in a power supply to protect the wiring. Large systems use a circuit breaker to protect the wiring from the dire effects of overload. Circuit breakers do double duty as main power on/off switches.

Outlets. One or more power outlets may be provided in the equipment for the use of peripheral subassemblies. Switched outlets are dead when the main power circuit breaker is off. Unswitched outlets are live when the main power circuit breaker is off, in order to provide power for trouble lights, a soldering iron, or other service aids.

Circuitry. A transformer isolates the circuit from the power line and converts the line potential to a value that is appropriate for the power supply. A rectifier converts the alternating current to direct current. The direct current obtained from a rectifier still has an ac component (called "ripple") that must be removed. Filter capacitors are used to smooth out the ac ripple from the rectifier. The output potential of an unregulated power supply will vary with line and load variations. A voltage regulator serves to hold the output potential from the power supply to a specified value. Considerable heat is dissipated in the power supply. Fans are used to cool the power supply and other electronics.

Circuit Boards. With the exception of the fans and the larger components in the power supply, and the panel-mounted controls and indicators, all of the electronic components are mounted on circuit boards. The conducting traces on the earliest circuit boards were actually printed on the board, using a conducting ink, and the boards were called "printed circuit boards" or "PCBs." Modern circuit boards are copper clad; and the copper is etched away except for the conductive traces, so the boards are more properly called "etched circuit boards," but the initials PCB are still used.

PART TWO
A HANDBOOK OF STYLE

> . . . to return back to the primitive purity, and shortness, when men delivered so many things, almost in an equal number of words. They have exacted from all their members a close, naked, natural way of speaking; positive expressions; clear senses; a native easiness: bringing all things as near the mathematical plainness as they can; and preferring the language of artisans, countrymen, and merchants, before that of wits or scholars.
>
> Thomas Sprat, *Early History of the Royal Society*, 1667

First impressions are notoriously stubborn. That is why it is extremely important that a company's technical publications—in which the customer is introduced to the product, and by extension to the producer—project the proper image. Well-written and well-edited manuals are characteristic of a company that has survived its formative years and is here to stay. Consistency of capitalization, spelling, and hyphenation; agreement of verbs and subjects; and close attention to such matters as beginning and ending quotation marks and parentheses, the number of ellipsis points, whether numbers are given as figures or written out, and many other details of style are the mark of maturity in publications.

We shall assume that you already know how to write, or you would not be pursuing a career in technical writing. If you do not know how to write, by all means read (nay, memorize) Strunk and White's excellent *Elements of Style* (Macmillan, 1979) right away; time's a-wasting. If you know everything that is in this little (eighty-five-page) book, you will know more than the majority of writers earning their living by the pen today.

The late Alfred North Whitehead characterized style as "the most austere of mental qualities . . . an aesthetic sense based on admiration for the direct attainment of a foreseen end; simply, and without waste."

Naturally, there is little to admire if you fail to attain your purpose; but it is nice if you can reach your goal without a great deal of splashing about. Willard Strunk, Jr., observed in *The Elements of Style* that a sentence should have no unnecessary words for the same reason that a drawing should have no unnecessary lines and a machine no unnecessary parts. This is excellent advice.

Whitehead viewed style as the "ultimate morality of the mind. With style," he said, "you obtain your end and nothing but your end, and with calculable effect. Your power is increased, because if your mind is not distracted by irrelevancies you are more likely to achieve your goal. Style," he said, "is the exclusive privilege of the expert and always the product of special study."

The canons of usage are the same for technical writing as for any other kind of writing, except for some important differences in the use of numbers and measurement (refer to Chapter 9). Technical writing follows newspaper practice in spelling out only the numbers under 10 rather than those under 100. Measurements are preferably expressed in the International System of Units (SI, for Système International). When so expressed, figures are used in place of words for the numbers, and symbols are used for the units. Other differences that distinguish technical writing from other writing, such as the prevalence of unit modifiers, are mainly a matter of degree.

Before you can begin to write, you will need a desk and a chair, some paper, a pencil or a pen, and some good reference books. After that, you can begin thinking about a wastebasket and maybe a word processor. This handbook is intended to supplement, not supplant, such standard reference works as *The Chi-*

cago Manual of Style. The reference works in the following list will answer nearly every question of style that may occur in technical writing.

- *Webster's New Collegiate Dictionary*, G. & C. Merriam Co., 1973. The *Collegiate* dictionary is revised more frequently than the unabridged and hence represents the latest thinking of the editors. For this reason the *Collegiate*, not the unabridged, should be the authority on matters of spelling and the like. (For instance, *baseline*—an important word in technical literature—is given as *base line* in *Webster's Third New International Dictionary* [1971] and as one word in the *Collegiate*.)
- *Webster's Third New International Dictionary*, G. & C. Merriam Co., 1971. Consult the unabridged for terms not found in the Collegiate Dictionary (e.g., allen screw).

  Note that technical dictionaries, as a rule, are not as well edited as *Webster's* and should be used only as a last resort.
- *The Chicago Manual of Style*, The University of Chicago Press, 1982. Become familiar with *The Chicago Manual of Style*. This is more than the style book of a distinguished publishing house; it is the bible of the publishing industry and is cited by publishers more than any other style guide.
- *Metric Editorial Guide*, American National Metric Council, 1978. The *Metric Editorial Guide* is a small booklet that tells you everything you need to know about the metric system.

Note that the topics in Part II are arranged in alphabetical order for ease of reference.

Abbreviations, Symbols, and Units

Abbreviations were v. common in the days when people wrote in longhand, but since the invention of the typewriter, the tendency has been to spell things out. No longer do we open a letter with "Yr. esteemed favor of the 17th. inst. to hand . . ." and close with "Yr. v. humble and obdt. servt." The old style persists in certain scholarly latin abbreviations (e.g., i.e., etc.) and some conventions such as Mr. and Ms., but for the most part, words are fully spelled out in formal communications.

In technical writing, the trend is in the opposite direction. Abbreviations are typically used in preference to words for units of measurement when preceded by a figure, for many terms that have become more familiar in abbreviated form than when spelled out, and for jaw-breaking polysyllabic terms that are used repeatedly in text.

Abbreviations must be used with restraint, however. Bear in mind that regardless of how familiar an abbreviation may appear to you, there will always be readers who are encountering it for the first time; and a text that is heavily laden with unfamiliar abbreviations will not be read.

An abbreviation is a shortened form of a word or expression, formed by dropping one or more letters. There may be more than one abbreviation in common use for some English expressions.

A symbol is an abbreviation that has acquired international sanction. There is only one approved symbol for any given expression. Three kinds of symbols are commonly encountered in technical writing: symbols for parts of molecules (Na, Me, Ile), symbols for physical quantities (*E*, *I*, *R*), and symbols for units of measurement (V, A, Ω). Note that symbols for parts of molecules are invariably written with initial capitals and symbols for physical quantities are invariably italic.

Symbols for units of measurement are extremely important, since they are used freely in text, whereas symbols for physical quantities, for instance, are used only in equations, tables, and illustrations, where it is necessary to save space.

Some expressions such as *dichlorodiphenyltrichloroethane*, *lysergesäurediethylamid*, *videlicet*, *confer*, and *maister* would hardly be recognized by the general reader in their unabbreviated forms. Their abbreviations (DDT, LSD, viz., cf., Mr.) are used freely in text without definition. Latin abbreviations (i.e., e.g., etc.) are in this class. Metric units are another important class of abbreviations that are used freely in text without definition, but only when preceded by a figure.

Another class of abbreviations comprises those that require definition on first use. Abbreviations such as *CPU* and *CRT* are a real help when the jawbreakers that they represent are used repeatedly in text. Bear in mind, however, that many such abbreviations will not be familiar to your readers. Space is never a problem in running text. The general rule is: Avoid abbreviations that require definition unless there is a genuine benefit to the reader—that is, unless the expression is used repeatedly, not just once or twice.

A fourth class of abbreviations comprises those that are used in equations, figures, and tables to save space. These usually do not require definition, being either familiar to the reader or readily understood from context. Chemical symbols and the symbols for physical quantities are in this class.

Additional abbreviations must be created to keep pace with new technology, subject to the constraints mentioned above.

The following are some rules to keep in mind.

1. Do not abbreviate the names of units except when they are preceded by a figure, or in tables to save space. Say "several milliliters," not "several mL."
2. Do not use periods with abbreviations in figures and tables, nor in text unless the abbreviation happens to spell an English word (fig., in., no.). Periods are generally used with Latin abbreviations (e.g., i.e., etc.).
3. Symbols for units (mL, mm, etc.) are always Roman.
4. Symbols for physical quantities (l, w, h) are always italic.
5. When in doubt, spell it out.

Standard Abbreviations. The following list comprises three classes of abbreviations: abbreviations that may be used freely in text without definition, abbreviations that may be used freely in tables and illustrations but should be spelled out in text, and abbreviations that must be defined on first use. The following abbreviations are standard in technical writing. You will have to add to this list the abbreviations that are standard in the particular field or industry that you are writing about.

alternating current	ac	define on first use
American National Standard Code for Information Interchange	ASCII	define on first use
American Wire Gauge	AWG	define on first use
ampere	A	
analog to digital converter	ADC	define on first use
analytical reagent	AR	define on first use
ante meridiem	AM	(use small capitals)
atomic weight	at. wt	
beginners' all-purpose symbolic instruction code	BASIC	define on first use
boiling point	bp	
British thermal unit	Btu	define on first use
carriage return	CR	define on first use
cathode ray tube	CRT	define on first use
central processing unit	CPU	define on first use
chemically pure	CP	

common business-oriented language	COBOL	define on first use
cycles per second	Hz	
degree (plane angle)	°	
degree Celsius	°C	
degree Fahrenheit	(convert to Celsius)	
diameter	diam	spell out in text
digital-to-analog converter	DAC	define on first use
direct current	dc	define on first use
edited, edition	ed.	
Editor	Ed.	
equation	Eq	
equivalent weight	equiv wt	
erasable PROM	EPROM	define on first use
field effect transistor	FET	define on first use
formula translation language	FORTRAN	define on first use
foot (300 mm)	ft	
freezing point	fp	
gram	g	
height (symbol)	*h*	define on first use
hour (3600 s)	h	
inch (25.4 mm)	in.	spell out in text
infrared	IR	
input/output	I/O	define on first use
inside diameter	ID	define on first use
integrated circuit	IC	define on first use
kelvin	K (not °K)	
length (symbol)	*l*	define on first use
light-emitting diode	LED	define on first use
liter	L	
maximum	max	spell out in text
megabyte	Mb	define on first use
melting point	mp	
meter	m	
minimum	min	spell out in text
minute (60 s)	min	

molar	M	
molecular weight (symbol)	*M*	
month (2.628×10 s)	mo	spell out in text
National Bureau of Standards	NBS	define on first use
National Institutes of Health	NIH	define on first use
natural logarithm	ln	
not applicable	NA	spell out in text
nota bene	N.B.	
nuclear magnetic resonance	NMR	
number	no.	
ohm	Ω	
original equipment manufacturer	OEM	define on first use
outside diameter	OD	define on first use
parts per million	ppm	
pascal	Pa	
percent	%	
post meridiem	PM	(use small capitals)
pound (453.6 g)	lb	
pounds per square inch (7 kPa)	psi	
pounds per square inch, gage	psig	
printed circuit board	PCB	define on first use
programmed ROM	PROM	define on first use
radio frequency	RF	define on first use
random access memory	RAM	define on first use
read-only memory	ROM	define on first use
reference	ref	
revolutions per minute	RPM	define on first use
root mean square	rms	
second	s	
silicon controlled rectifier	SCR	define on first use
specific gravity	sp gr	define on first use
Système International d'Unites	SI	define on first use
temperature	temp	spell out in text
(Celsius, symbol)	*t*	spell out in text
(Kelvin, symbol)	*T*	spell out in text
test point	TP	define on first use

2^{10} (1024)	K	(see note)
2^{20}	M	
ultraviolet	UV	
United Kingdom	U.K.	(adjective only)
United States	U.S.	(adjective only)
volts ac	V ac	define on first use
volts dc	V dc	define on first use
weight	wt	
weight per weight	w/w	
width (symbol)	*w*	spell out in text
yard (914.4 mm)	yd	spell out in text
year (3.155 693 × 10 s)	yr	spell out in text

Note: The symbol *K* is used to denote the numerical quantity 1024 (2^{10}), which is commonly encountered in binary counting systems. It should be preceded by a numeral and a space. This symbol is not interchangeable with *k*, the SI prefix denoting × 1000.

New Abbreviations. When new abbreviations are needed (no standard abbreviation exits), follow the modern trend of eliminating periods and using all capital letters; e.g., *analog-to-digital converter: ADC* (not *A.D.C.* or *adc*).

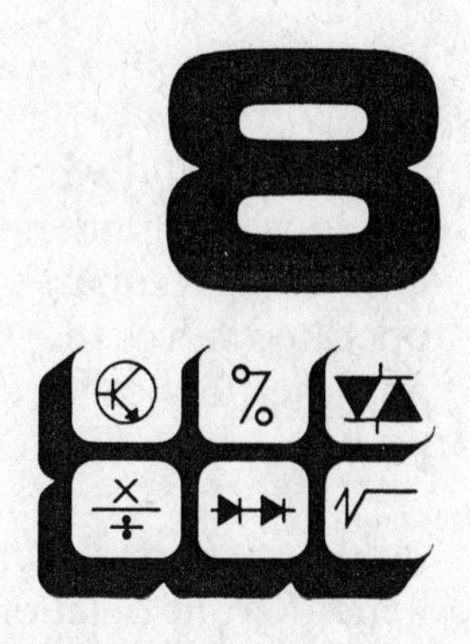

8 Bastard Enumeration and Other Topics

This chapter consists of a number of brief, highly opinionated essays on various topics, arranged alphabetically for easy reference.

□ BASTARD ENUMERATION

In a simple series such as "red, white, and blue" or "coffee, tea, or milk," a conjunction (*and* or *or*) is required before the last item. If this conjunction is omitted, a chaotic situation known as "bastard enumeration" results. This frequently happens when there is a series within a series, as in the following example written by a dedicated practitioner of the art of bastard enumeration.

> The CI controller interconnects the vacuum system, reagent gases, ionizer, and the gas chromatograph. This interface enables selection of gases, vacuum and ionizer control from pushbutton switches on the front of the vacuum system. Figure 9-2 illustrates which electrical control signals from the vacuum controller and the QEM provide selection of reagent gas, carrier gas, ionizer control, and plumbing evacuation. Check for defects such as broken or loose connections, improperly installed fittings, plugs and bent or heat damaged components.

To understand what went wrong in the foregoing example, it is necessary to understand what the fellow is trying to say. He is describing a subassembly that consists of a number of electrically operated valves that are used to control the application of vacuum and the flow of various gases in a system, and he is suggesting places to look for defects in the event of a malfunction.

The first sentence is all right as far as logic is concerned (and logic is really what we are concerned with here), but it could benefit from deletion of the word *the* in front of *gas chromatograph*. The *the* in front of *vacuum system* would then extend its influence to cover four parallel elements: *vacuum system, reagent gases, ionizer,* and *gas chromatograph.* This device, known as "parallel construction," is a powerful aid to comprehension of the structure of a writer's thoughts.

The second sentence fails because of a false parallelism suggested by the omission of a needed connective between *gases* and *vacuum*. The sentence seems to say that the subassembly enables selection of three elements: gases, vacuum, and ionizer control. What the writer intended was to say that the subassembly enables two elements: selection (of gases) and control (of vacuum and the ionizer).

The writer then proceeds to compound the confusion, in the guise of amplification. In the third sentence, we have a good example of a series within a series rendered chaotic by bastard enumeration. The main series consists of the nouns *selection, control,* and *evaporation;* and the inner series consists of *reagent gas* and *carrier gas*. The conjunction is needed in the inner series. What was intended in the third sentence was that *provide* should govern three elements: *selection, control,* and *evacuation,* and that *selection* should govern only the gases.

Finally, in the fourth sentence, the writer seems to be telling the operator to check for plugs, which may be a good idea but is not what the writer intended. The phrase *fittings, plugs* should read *fittings and plugs*.

So there, in one paragraph of four sentences, we have three examples of bastard enumeration—possibly not a record but certainly a sterling example of what can happen when a truly dedicated bastard enumerator is out of control.

With the aid of this analysis (and one or two topics discussed elsewhere in this work), let us recast the offending paragraph so that the operator may have some inkling of what is going on and what he or she ought to do.

> The CI controller interconnects the vacuum system, reagent gases, ionizer, and gas chromatograph. This interface permits selection of gases and control of vacuum and the ionizer by means of push-button switches on the front of the vacuum system. Figure 9-2 shows which control signals from the vacuum controller and QEM provide reagent and carrier gas selection, ionizer control, and plumbing evacuation. Check for such defects as broken or loose connections, improperly installed fittings and plugs, and bent or heat-damaged components.

□ CHEMISTRY

The amount of a chemical substance is expressed in moles, abbreviated *mol*. The volume of a liquid is commonly expressed in liters or milliliters. The international symbol for liter is *l*, but since on a typewriter it looks so much like a 1, the American National Metric Council recommends use of a capital *L*. The ratio of volume to quantity may be expressed in normality (*N*) or molarity (*M*).

Nomenclature of Organic Compounds. Two conventions frequently overlooked in writing the names of organic compounds are word division and the use of italics.

Word Division. The names of organic compounds are printed as one word, regardless of length, for example, *decafluorophenylphosphine*. The exception to this rule is that classes of compounds (acids, lactones, sulfones, sulfoxides, etc.) are divided—for example, *acetic acid*.

Italics. Configuration prefixes (*cis*-, *trans*-, *sec*-, *tert*-, etc.) and locants (*N*, *S*, *O*, etc.) are set in italic type. The positional prefixes, *o*-, *m*-, and *p*-, should be replaced by 1,2; 1,3; and 1,4 respectively.

☐ DANGLING CONSTRUCTIONS

"While absent on February 24, and possibly February 25, Cor Laan will be in charge and will be authorized to act in all matters normally brought to my attention." This gem is from an actual interoffice memorandum. Such ambiguities, often called dangling constructions, are best avoided, not because they are misleading but simply because they are not acceptable to the majority of educated readers. Being often ludicrous, one must strive to avoid dangling constructions.

☐ EMPHASIS

Typesetting offers a number of devices to give special emphasis to words that, for one reason or another, require special treatment. Do not use "quotation marks" for emphasis. The use of Initial Capitals to stress Important Words is an old-fashioned device now chiefly used for satiric effect. Underlining is an editing mark used to specify italic type; words that are underlined in manuscript will be set in italics. ALL CAPITALS slows reading down 20 to 50 percent. Italics are 10 to 50 percent less legible than Roman letters. Use capitals or italics when you want to slow the reader down. Otherwise, use bold type for emphasis. (We want to slow the reader down when hasty reading of a sentence or paragraph could result in the misreading of a warning or caution, for instance. In technical writing, we do not want the reader to skim safety warnings and the like.)

Foreign words that have not been completely assimilated into English should be set in italics. Use italics to introduce a new term that may not be familiar to the reader.

Panel legends (words silkscreened on the panel) should be set in the same kind of letters that are used on the panel. This is why panel legends nearly always appear in capital letters in text.

☐ ENUMERATION

In a simple series that is run into the text, either figures (1) or italic letters (*a*) may be used to indicate the items. There seems to be little reason for choosing one over the other. The Old Tech Writer says, "I always use figures in enumerations when I'm word processing, because they take fewer keystrokes than italic letters."

☐ HEADLINES

Headlines are added to the text for the convenience of the reader, but they are part of the graphics, not part of the text; therefore, the text should not refer to the headlines but should stand on its own feet. When a paragraph following a headline starts with "This is . . .," or words to that effect, amend it to repeat the substantive in the headline. (Note that this paragraph starts with *Headlines,* not with *These.*)

☐ HYPHENS

> One must regard the hyphen as a blemish to be avoided whenever possible.
>
> Winston Churchill

> Hyphening is more an editor's worry than a writer's.
>
> Porter G. Perrin, *Writer's Guide and Index to English*, Scott Foresman, New York

> If you take hyphens seriously, you will surely go mad.
>
> John Benbow, *Manuscript and Proof* (the style book of Oxford University Press, New York)

According to the University of Chicago Press style manual, "nine out of ten spelling questions that arise in writing or editing are probably concerned with compound words." The *United States Government Printing Office Style Manual* devotes more than fifty pages to the subject. Obviously, this is not a topic that can be explained by a few simple rules.

Compounds may be classed as permanent (*baseline*) or temporary (*heat-damaged*). Permanent compounds are those that have attained sufficient currency to be listed as vocabulary entries in a dictionary; you can look them up. The temporary compounds are the ones that cause all the trouble.

Prefixes and Suffixes. Do not hyphenate prefixes unless the root begins with a capital letter (e.g., *non-SI*) or the combination spells a different word (*un-ionized*). This rule holds even when the prefix ends and the root begins with the same vowel (*cooperate, preexisting, reentry*).

Whereas unit modifiers (compounds used as adjectives) typically require a hyphen to avoid ambiguity or absurdity, prefixes and suffixes are generally set solid with the root. Sometimes it is not immediately clear whether a given modifier is a derivative (a word derived by the addition of a prefix or suffix) or a unit modifier. The following affixes are commonly used in other senses as separate words. Do not use a hyphen when forming derivatives with these affixes.

Eight Prefixes That Resemble Words. Beware of the following prefixes, which have different meanings when used as separate words. Do not hyphenate

extra	over	pro	ultra
out	post	super	under.

Two Suffixes That Resemble Words. *Like* and *fold* have different meanings when used as suffixes. Do not hyphenate.

Fifteen Words That Resemble Prefixes. The following words retain their meanings when used as a part of a unit modifier. They are not prefixes and do require a hyphen.

all-	higher-	little-
best-	highest-	lower-
better-	ill-	quasi-
cross-	larger-	well-
half-	lesser-	wild-

Unit Modifiers. Technical writing abounds in unit modifiers. Except when used in the predicate, unit modifiers typically require hyphens to prevent misreading. Exercise restraint; be sure that the hyphen is really needed to prevent mispronunciation or avoid ambiguity. The following paragraphs are included as a general guide only. The final decision has to be governed by the sense.

Rule: Use a hyphen to form unit modifiers except when there is absolutely no possibility of misleading the reader. If the reader can stumble, even for an instant, then a hyphen is required. (Often the question is not one of sense but of sobriety, for example, a "camel's hair brush" versus a "camel's-hair brush.") In general, use a hyphen if a compound multiplier comes before the noun (*10-mL tubes*).

Adverbs. As Fowler remarked,* it is the business of an adverb to modify the word that is next to it; hence, there is little possibility of confusion when an adverb is used as the first element of a unit modifier. Do not hyphenate. Example: *a totally awesome experience*.

Chemical Compounds. The names of chemical compounds are not hyphenated when used as unit modifiers. Example: *sodium chloride solution*.

Colors. If both elements of a unit modifier are colors, no hyphen is required.

Fractions. Fractions used as unit modifiers require hyphens. Example: *two-thirds full*.

Letters and Numbers. Letters and numbers used as the first element in a compound multiplier require a hyphen. Letters and numbers used as the second element do not. Be particularly aware when using *odd* as the second element in a unit modifier. "Twenty-odd technical writers" is not the same as "twenty odd technical writers." Examples: *T-Model Ford* but *Model T Ford*.

Phrases. Hyphenate English phrases used as unit modifiers. Foreign phrases require no such hyphenation, being sufficiently

*Fowler, H. W. *A Dictionary of Modern English Usage.* 2d. ed. revised by Sir Ernest Gowers. Oxford: Clarendon Press, 1965.

united by the fact that they are foreign. Examples: *blue-ribbon jury* but *veal cordon bleu.*

Proper Names. Unless they are hyphenated to begin with, proper names require no hyphenation when used as modifiers, being united by their capitalization. Example: *General Electric.*

Scientific and Technical Terms. Scientific and technical terms require no hyphenation when used as unit modifiers. Example: *stainless steel tubing.*

□ ITALICS

Typical uses of italic type in technical manuals include *continued* and like phrases in brackets, distinctive treatment of words (for emphasis, foreign words, special terms), algebraic quantities, to indicate illustrations in indexes, letters in enumerations, mathematical expressions, titles of published works, and run-in sideheads.

□ JARGON

Abbreviated terms like *boot* (for *bootstrap*) and *scope* (for *oscilloscope*) should be spelled out for the benefit of the uninitiated. Even if your audience can reasonably be expected to know the terms, there are people born every minute who will someday be reading them for the first time. Give them some consideration.

□ LEGEND COPY FOR ILLUSTRATIONS

Legend copy for an illustration may consist of a caption, a legend, or both. (In technical writing, as at the University of Chicago Press, captions and legends are "distinguished elegantly the one from the other.") A caption tells what is in the picture ("Rear view of the J21D, showing electrical connections"). A caption is not a grammatical sentence. A legend consists of one or more grammatical sentences and should be given sentence capitalization and punctuation. If a figure has a caption and a legend both, use a period after the caption.

□ MATHEMATICS

Mathematical copy is a pain in the neck for writers and typesetters and thus is often considered "penalty" copy. Here are a few pointers.

Identification of Symbols. Symbols used in equations must be identified.

After Equations. Frequently, the symbols used in an equation will be identified immediately following the equation wherein they are first used. If many symbols are to be identified, the definitions themselves should be displayed and aligned on the equals signs, preceded by the word *where*, typed flush left:

where

μ = mean of the population
$\bar{\chi}$ = arithmetic mean of the sample
σ = standard deviation of the population

If only a few symbols are to be identified, they may be used within the text as follows:

> where r is the radius in millimeters, ω is the angular velocity in radians per second, and g is the standard acceleration of gravity.

In Monographs. The formal listing of symbols used throughout a monograph is usually part of the front matter and immediately precedes the first section of the text. It is normally headed

```
                    Notation
ν     viscosity
λ     wavelength
ω     angular velocity in radians per second
π     3.1416
ρ     density in grams per milliliter
σ     standard deviation
      etc.
```

Figure 21. Formal listing of symbols.

"Notation." Each entry consists of a symbol and its identification (see Figure 21).

Appositives. Symbols are frequently introduced in text in apposition to the words that they stand for. When so introduced, they are usually what is known as "close" (restricting or complementary) appositives and are used without punctuation.

> The force exerted by a unit mass rotating at a velocity ω and a distance r about a fixed point is given by
>
> $$\text{force field} = \omega^2 r$$

Occasionally, a previously identified symbol may be regarded as a "loose" (nonrestricting or supplementary) appositive and be set off with commas or parentheses.

> The angular velocity, ω, must commonly be computed from the rotational speed in rpm.

Operational Signs. Operational signs are words and deserve space fore and aft. Write "$a \ = \ b$," not "$a{=}b$."

Punctuation. If a sentence ends in a mathematical expression, put a period after it.

> Hence it is apparent that
>
> $$abc = xyz.$$

Typography. Typesetters should know that letters used to represent mathematical quantities are usually italic. Nevertheless, it is a sensible precaution to underline such letters whenever they appear in a manuscript. If Greek letters and special signs are drawn by hand, they must be identified in the margin.

Equations. Center displayed equations and leave extra space above and below. Break long equations, if necessary, before an operational sign (preferably an equals sign). The operational sign must be carried to the next line.

Use the slash (/) when fractions are typed in line in the text. Use a horizontal fraction bar in displayed equations.

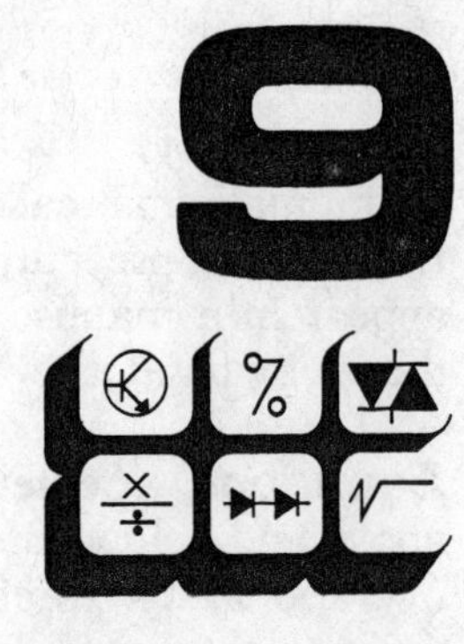

Numbers and Measurement

Technical writing differs from general writing chiefly in the treatment of physical quantities, and this in two ways: Figures are often used where general writing would require words, and the units following the figures are nearly always abbreviated.

□ NUMBERS

The following rules apply to numbers in text.

Rule 1: Do not begin a sentence with a figure.

Rule 2: Use figures instead of words in the following situations, except when such use would conflict with Rule 1.

a. serial numbers, such as dates and references (Section 1, Figure 2, Table 3)
b. numbers used in computation (add 1, subtract 2, multiply by 3)
c. numbers in the same category within a single paragraph, if one of them must be a figure

"In standard shift operation, 5 or 6 photographs can be taken per plate. For equilibrium interference work requiring many exposures, the swinging-gate assembly is used in conjunction with the short-shift mechanism to permit as many as 126 exposures per plate."

Here, since 126 must be a figure (refer to Rule 4), it was necessary for the writer to write "5 or 6" instead of "five or six."

d. measurements in the International System of Units (the "metric" system), or when followed by an abbreviation or symbol, except when separated from the following unit, abbreviation, or symbol by one or more words

1 mm, 2 g, 3 L, 4 s, etc.; but "one or more milliliters . . ."

e. compound fractions, decimal fractions, and percentages

Note: Always use the "%" sign in technical writing except at the beginning of a sentence. ("Twenty-seven percent of the cost was guaranteed . . .")

Rule 3: Use words instead of figures in the following situations, except when such use would violate Rule 2.

a. When quantity and measurement occur in the same expression, use words for the quantity, unless the quantity is very large ("Load the sample injector with seventy-two 10-nmol samples . . .").

b. round numbers and approximations

Rule 4: In situations not covered by Rules 1, 2, and 3, use words for cardinal numbers under ten, figures for ten and over. (The rule for ordinal numbers is the same except that figures start with eleven—tenth, 11th, etc.)

Decimal Point. Always use a zero before the decimal point for values less than one.

Fractions. In general, decimal fractions are better than common fractions (1.5, not 1 1/2), except for very rough approximations where the use of decimals might imply more precision than is warranted (1/2 turn).

Grouping. Since the comma is used as a decimal point in many countries, a comma should not be used to separate groups of digits. Use a space. Write four-digit numbers solid, except in tables.

□ MEASUREMENT

If one thing can be said to characterize technical writing, it is measurement. Technical writing abounds in expressions of length, mass, volume, and other dimensional properties, and it behooves the technical writer to become proficient in their use.

For the last 200 years, the world has been moving in the direction of an international system of units for measurement. This international system (SI, for Système International, commonly called the metric system) now seems destined for adoption by the United States. Scientists have been using it for years.

Metricizing, for the writer, consists of three processes: conversion, rounding, and—occasionally—rationalizing. Conversion consists of multiplying a nonmetric value by an appropriate conversion factor to obtain the metric equivalent. Rounding consists of reducing the number of significant digits in the converted value so that the converted value does not imply greater precision than is warranted. Rationalizing consists of substituting a round number in one system for a round number in the other so as to convey an equivalent connotation. The rules for conversion and rounding are given in the *Metric Practice Guide E380–76*, published by the American Society for Testing and Materials.

Metric practice should conform to the recommendations of the American National Metric Council as presented in the latest edition of the *Metric Editorial Guide*. The following text is by way of amplification of the precepts contained therein.

Compound Units. Compound units are units that are formed by combining base units, using the operations of multiplication and division, or by adding exponents.

Area and Volume. Use an exponent with the symbol to express area or volume (cm^3, not *cc*).

Products. When writing the name of a compound unit that is a product (e.g., *newton meter*), leave a space between the words. Do not use a hyphen.

Quotients. When writing the name of a compound unit that is a quotient, use *per* between the words (*grams per liter*). Use a slash (also known as a diagonal, shilling mark, slant, solidus, or virgule) to form the symbol. Do not use more than one slash per expression; *a/b/c* is mathematically ambiguous. For example, 1/2/3 could mean $^1/_2$ divided by 3 = $^1/_6$ or 1 divided by $^2/_3$ = $1^1/_2$. $1 \times 2^{-1} \times 3^{-1}$ unambiguously means $1 \times {}^1/_2 \times {}^1/_3 = {}^1/_6$.

Note: A dandy unambiguous way to express quotients is with negative exponents: $\mathrm{rad}^{2 \cdot -1}$ = rad^2/s.

Conversion and Rounding. Measurements given in traditional units (inches, pounds) should be converted to the International System (SI) and rounded in such a way that the original precision is neither sacrificed nor exaggerated. Use the following procedure.

1. Convert feet and inches to inches; pounds and ounces to ounces.
2. Multiply by the appropriate SI conversion factor.
3. If the first significant digit of the converted value is equal to or larger than the first significant digit of the original value, round the results to the same number of significant digits as in the original value. For example, 2 in. × 25.4 = 50.8 mm; round to 50 mm. If the first significant digit of the metric value is smaller than the first significant digit of the original value, round the results to one more significant digit than in the original. For example, 30 psig × 6.895 = 206.85 kPa. Round to 210 kPa, not 200.

These simplified rules are not a substitute for common sense. It is incumbent upon the writer to make an educated guess as to the degree of precision intended by the original measurement and to express the metric equivalent accordingly. For instance, a measurement given as 10.25 in. may actually mean $10^1/_4$ in., with a probable tolerance of $\pm{}^1/_8$ in., in which case, 260 mm would be a more reasonable conversion than 260.4 mm, as would be required by the simplified rule given previously. Proper application of the rules for rounding requires some knowledge of how measurements are ordinarily made.

Selected Conversion Factors.

To convert from	*to*	*multiply by*
Btu/hour	watts	0.293
inches	millimeters	25.4
feet	millimeters	304.8
yards	millimeters	914.4
pounds	grams	453.6
°f	°C (subtract 32 and multiply by 5/9)	
rpm	radians per second	0.104 7
psi	kilopascals	6.894 757[a]
torr	pascals	133.322 36

[a]Use 7 for ordinary work. (By "ordinary work," I mean work not requiring an extraordinary degree of precision—like the National Bureau of Standards.)

Derived Units. Other units in the SI system are derived from the basic units by multiplication or division. In writing the expression for a derived unit, use a raised period to indicate multiplication ($a \cdot b$) and a slash for division. Multiply top and bottom of quotients by an appropriate factor to eliminate prefixes in the denominator. (Exception: For historical reasons, the kilogram is a base unit, and *kg* should be used in the denominator rather than *g*.)

Dual Dimensioning. Some companies, during the transition from traditional (by "traditional," I mean measurements expressed in feet and inches, pounds and ounces) to metric measurement, will use dual dimensioning on their engineering drawings and in their technical literature, for example:

Length	50 mm (2 in.)
Weight	1 kg (2.2 lb)
Pressure	140 kPa (20 psig)

Measurements that have traditionally been expressed in metric units (time, amount of substance, electric current, frequency, power, electromotive force, resistance) will not require dual dimensioning. For example:

Elapsed time	60	s
Current	25	A
Sample size	10	nmol
Frequency	60	Hz
Power	100	W
Potential	110	V ac
Resistance	300	Ω

Do not apply dual dimensioning to nominal dimensions, that is, to dimensions that merely name the item (for instance, there is nothing "one inch" about one-inch pipe), nor to time or temperature measurements. Say "35-mm film," not "35-mm ($1^{3}/_{8}$ in.) film"; "0°C," not "0°C (32°f)"; "1.5 h," not "1.5 h (5.4 ks)."

Do not use dual dimensioning in an equation. If non-SI units are used in an equation, either restate the equation in SI units or add a note specifying the appropriate conversion factor. When dual dimensioning is used in an illustration, observe the conventions depicted in Figure 22.

Figure 22. Dual dimensioning in illustrations.

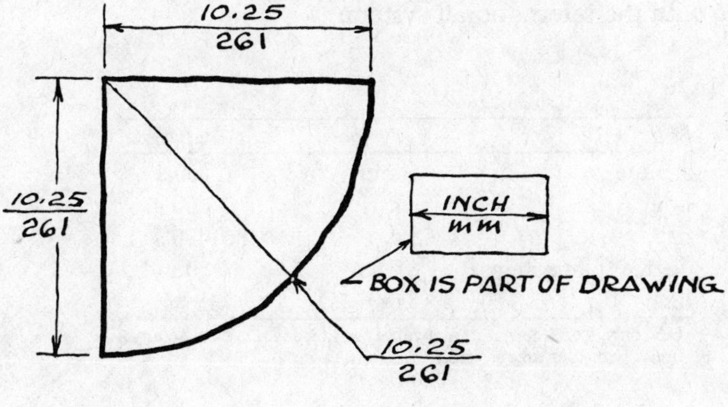

Grammatical Number. Names of units are singular for all values between minus one and plus one except zero, which is plural (zero degrees). All other values are plural. The names of the units form their plurals in the usual way. The symbols are always singular, for example: −2 degrees, −1 degree, −0.5 degree, 0 degrees, 0.5 degree, 1 degree, 2 degrees.

Mass and Weight. Weight is the force developed by mass when it is subjected to a particular acceleration. Nevertheless, in commercial and everyday use in the United States, the word *weight* nearly always means mass. Technical writing must be careful to distinguish between the two in any context in which the meaning is not completely clear.

Nominal Dimensions. A nominal dimension names the item and hence does not require metricizing. *One-inch pipe, two-by-four studs, quarter-inch recording tape, yardsticks, quart bottles, three-by-five cards, eight-by-ten prints,* and *8½ × 11 paper* are examples of nominal dimensions.

Non-SI Units. A number of non-SI units are commonly used with the metric system. Some are accepted, some have temporary sanction, and some are to be avoided wherever possible.

Accepted Units. The following non-SI units are accepted for use with the international system.

Unit	*Symbol*	*SI Equivalent*
minute	min	60 s
hour	h	3 600 s
day	d	86 400 s
degree (plane angle)	°	π/180 rad
liter	L[a]	10^{-3} m³

[a] U.S. only. The international symbol for liter is *l*. Do not use any prefix with liter except "milli-."

Temporary Units. The following non-SI unit will be phased out in a future resolution of the International Committee on Weights and Measures.

Unit	*Symbol*	*SI Equivalent*
atmosphere	atm	101 325 Pa

To Be Avoided. The following non-SI units should be avoided, per solution of the General Conference on Weights and Measures (CGPM) in 1960.

Unit	*SI Equivalent*
torr	101325/760 Pa
micron	10^{-6} m (= 1 μm)
lambda	10^{-9} m^3 (= 1 mm^3)

Prefixes. Quantities should be expressed using no more than three digits to the left of the decimal point. Table 3 provides prefixes approved for this purpose; however, it is generally best to pick a prefix and stick with it throughout a discussion, table, or illustration unless the discrepancy in dimensions is really gross (e.g., 30 m of 0.5-mm ID tubing). Do not use double prefixes; say "nanometer," not "millimicrometer." Note that the first five prefix symbols in Table 3 are capital letters. The rest are lower case.

When quantities are squared or cubed, many more digits to the left of the decimal may be generated. It is not possible to reduce these to fewer than four digits by using the prefixes in Table 3. For this reason, the prefixes listed in Table 4 are approved for expressing areas and volumes. Consider, for instance, the dimension 200 mm. Neither the square nor the cube of this value can be expressed in fewer than five digits, nor is it any help to state the dimension in meters, since the object is to express the quantity as a number between 1 and 999; 0.2^2 is 0.04, and 0.2^3 is 0.008; but if the dimension is stated in centimeters (1 cm = 0.01 m), the area can be expressed as 20^2 = 400 cm^2; and if the dimension is stated in decimeters (1 dm = 0.1 m), both the square

Table 3. SI Prefixes.

Multiplier	*Prefix*	*Symbol*
10^{18}	exa	E
10^{15}	peta	P
10^{12}	tera	T
10^{9}	giga	G
10^{6}	mega	M
10^{3}	kilo	k
10^{-3}	milli	m
10^{-6}	micro	μ
10^{-9}	nano	n
10^{-12}	pico	p
10^{-15}	femto	f
10^{-18}	atto	a

Table 4. Supplementary Prefixes.

Multiplier	*Prefix*	*Symbol*
10^{2}	hecto	h
10	deka	da
10^{-1}	deci	d
10^{-2}	centi	c

and the cube can be expressed as one-digit numbers ($2^2 = 4$; $2^3 = 8$).

Pressure. The SI unit of pressure is the pascal (rhymes with *rascal*). The kilopascal (= 1000 pascals) is recommended for all pressure measurements in excess of atmospheric pressure (which is about 100 pascals). Vacuum measurements may be expressed in pascals (Pa) or millipascals (mPa).

To convert measurements given in pounds per square inch (psi or psig) to kilopascals, multiply the psi value by 7; for example, 20 psi = 140 kPa. (This is not exact but is close enough unless you are reporting some extraordinarily precise measurements.)

Absolute Pressure. Absolute pressure is the pressure measured by a gauge plus the atmospheric pressure. For instance, if a gauge measures 110 kPa and the barometric pressure is 100 kPa, the absolute pressure is 210 kPa. In metricizing, assume that pressures expressed in pounds per square inch (psi or psig) are gauge pressures unless otherwise indicated; absolute pressures are seldom encountered in technical writing. If absolute pressure is given (psia), multiply by 7 and add the word *absolute* (not abbreviated). For instance, 30 psia = 210 kPa absolute.

Approximate Conversion Factors. Pressure measurements are usually not very precise. If you can tolerate an uncertainty of approximately 2 percent, you can save time by using the approximate conversion factors shown in the accompanying table.

To convert from	*to*	*multiply by*
atmospheres	kPa	100
inches of mercury	kPa	3.4
torr	Pa	133
psi or psig	kPa	7

Punctuation. Do not use a period after a metric symbol except at the end of a sentence. To facilitate the reading of large numbers, use spaces instead of commas. This style helps avoid confusion caused by the European use of commas to express decimal points. Do not use a space in four-digit numbers, except in tables.

Rationalizing. Rationalizing consists of translating a noncritical value into a popular, standard metric equivalent. For example, a 2-in. specimen becomes 50.8 mm by conversion, 51 mm by rounding, and 50 mm by rationalization. There are no rules for rationalizing, because rationalizing is an art, not a science. The writer must make a judgment as to the degree of precision intended by the original dimension and select an appropriate equivalent in the metric system. For instance, if you are translating a western story into French, it would not do to call a ten-gallon hat a 37.85412-litre chapeau, or even thirty-eight litres. You might get by with forty litres.

Spacing. In general, leave a space between the number and the unit, except for °C, which is typed solid with the number (*20°C*, not *20° C*). When a quantity is used as a unit modifier, use a hyphen between the number and the symbol (35-mm film).

Temperature. The SI unit of temperature is the kelvin (not *degree kelvin*), and its symbol is *K* (not *°K*). One kelvin equals 1/100 of the temperature range of liquid water, from freezing to boiling. Absolute temperatures stated in kelvins are measured from absolute zero—the theoretical temperature at which all molecular motion ceases. The more commonly used Celsius scale starts at the freezing point of water, 273 K. One degree Celsius is exactly equal to one Kelvin. The symbol is typed solid (no space) with the number (20°C).

Time. The SI unit of time is the second (s), which should be used whenever calculations are involved. If calculations are not involved, minutes, hours, days, and so on may be used.

Selected Units. The following units and their symbols should be committed to memory.

Unit	*Symbol*
ampere	A
hertz	Hz
kilogram	kg
meter	m
mole	mol
ohm	Ω
pascal	Pa
radian	rad
second	s
volt	V
watt	W

The Passive Voice and Other Problems

This chapter contains more opinionated essays, arranged alphabetically in the manner of Chapter 8.

□ THE PASSIVE VOICE

Textbooks tell us to shun the passive voice, but the passive voice has its place. Use the passive when the object of an action is of interest and the agent is relatively unimportant. For instance, it is appropriate to say "The president was shot" rather than "Somebody shot the president." Compulsive overuse of the passive, however, can lead to awkward sentences.

□ PARENTHETICAL EXPRESSIONS

Parenthetical expressions are expressions that can be deleted from a sentence without destroying the sense of the sentence; they provide additional information. Parenthetical expressions must be separated from the rest of the sentence by commas, dashes, or parentheses. The important thing to remember about parenthetical expressions is that the punctuation separating such expressions from the rest of the sentence usually comes in pairs. Parentheses always come in pairs. Commas or dashes are used in pairs except when the parenthetical expression comes at the end of the sentence. Omission of one of the required commas (or dashes) is a very common error.

Here is a foolproof method for determining whether an expression is parenthetical or not: Try enclosing it in parentheses. If the sentence still makes sense, the expression is parenthetical and may be enclosed between commas, parentheses, or dashes as you see fit. If the sentence looks stupid, the expression is not parenthetical and should not be enclosed between punctuation. Examples:

> This house (which was built by Jack) is full of malt.

The enclosed expression is parenthetical.

> This is the house (that Jack built).

The enclosed expression is obviously not parenthetical. Refer to *That* and *Which* under "Usage" in the following pages for more on parenthetical expressions.

□ PUNCTUATION

Punctuation is used in technical writing where needed (1) for clarity and (2) to meet the expectations of educated readers. Unnecessary punctuation should be omitted. Follow the recommendations of *The Elements of Style* (Strunk and White, Macmillan, 1959).

The rules of punctuation are not secret; they were drilled into us through years of grade school and high school; we had a second dose of them in college; and we can find them summarized in many standard reference books, including those listed at the end of this section. Furthermore, many of the rules are neither hard nor fast. The English language permits great latitude in punctuation. Yet technical writing abounds in the most elementary of blunders. This section highlights some of the punctuation problems that occur repeatedly in technical writing.

Commas. A few rules regarding commas are obligatory; aside from these, the use of commas is mainly a matter of good judgment, with ease of reading as the end in view.

Compound Sentences. A comma is required before the conjunction in a compound sentence, unless the sentence is very short ("You wash and I'll dry").

Parenthetical Expressions. Commas are used in pairs to enclose parenthetical expressions (refer to the previous section "Parenthetical Expressions"). Omission of one or both of the commas is a very common error in technical writing.

Series. Usage is divided on the question of whether a comma is needed before the *and* in a series such as "red, white, and blue." One school of thought maintains that since the comma between *red* and *white* represents a missing conjunction, no comma is needed between *white* and *blue,* where the conjunction is present. Consider, however, the series *triglycerides, purines and pyrimidines and proteins*. If you are a biochemist, you might know that purines and pyrimidines constitute one class of substances. If you are not, you might be confused by the two *and*s in the series. Rather than entertain a separate rule for such situations, most style books recommend a comma after each item in a series except the last.

Semicolons. We learned in school that splicing independent clauses together with a comma is wrong but that if you use a semicolon it's all right. Actually, there is a little more to it than that. The semicolon, being a stronger punctuation mark than a comma but not as strong as a period, is occasionally used to join together two closely related clauses. The key phrase here is *closely related*. If the clauses are not closely related, the appropriate punctuation mark is a period.

The principal objection to this use of the semicolon is that it results in long sentences; and as everybody knows, long sentences are more difficult to comprehend than short ones. Particularly in technical writing, one feels that the reader is in trouble most of the time and needs all the help he can get. One way that we can help is by keeping our sentences short.

Periods. The sin of omission (of periods) is strangely absent in technical writing. Indeed, so strongly ingrained is the period habit that we are continually placing periods where they are not needed. A page that is peppered with periods presents a pocked appearance, and one of the last editorial chores preparatory to printing is often going over the camera-ready art and whiting out gratuitous periods—or opaqueing them on the negative or scratching them off the plate. You never get them all.

The modern trend is toward elimination of periods in abbreviations, although educated people are still inclined to regard the omission of periods in Latin abbreviations as an oversight. Abbreviations that spell an English word (*fig., in., no.*) require periods.

Omit periods after headlines (except run-in heads), titles, simple captions, and display lines. Any punctuation following a computer commmand is particularly hazardous; for instance, a period following a numerical entry could be interpreted by the computer as a floating decimal.

Items in a vertical list normally do not require periods; however, if one of the items in the list comprises more than one sentence, then periods may be needed throughout the list for consistency of appearance. If a table title is followed by a headnote, a period is needed after the title. If a figure has both a caption and a legend, a period is needed after the caption.

Quotation Marks. The best time to use quotation marks is when you are actually quoting somebody. Do not use quotation marks to set off a nonstandard expression. If an expression is appropriate, use it without apology; if it is not, don't use it.

Suggested Reading. Detailed information on punctuation may be found in many standard reference works, including the following.

- Bernstein, T. M. *The Careful Writer: A Modern Guide to English Usage;* Atheneum, New York, 1965.
- Fowler, H.W. *A Dictionary of Modern English Usage,* 2nd ed. revised by Sir Ernest Gowers; Oxford: Clarendon Press, 1965.
- *The Chicago Manual of Style,* 13th ed., rev; The University of Chicago Press, Chicago, 1982.
- Perrin, P. G. *Writer's Guide and Index to English,* 4th ed.; Glenview, Ill.: Scott, Foresman & Co., 1968.
- Strunk, W. S., Jr., and White, E. B., *The Elements of Style,* 3rd ed.; Macmillan: New York, 1979.
- *Webster's New Collegiate Dictionary;* G. & C. Merriam Co., Springfield, Mass., 1977, pp. 1520 ff.

□ REFERENCES

Use the Harvard (name and date) system when compiling a list of references (see Figure 23). Minimize stops and commas. Do not use *ibid.* and like abbreviations.

```
Braun, A., Brown, B. & Le Brun, C. (1981) Journal, 11,
111-113.
```

Figure 23. Typical bibliographical entry. Note the following features: (1) initials follow surnames, (2) use of ampersand (&) instead of "and" (no comma before), (3) year in parentheses immediately following names, (4) journal title (abbreviated) and volume number in italics, and (5) first and last pages.

□ SPELLING

For spelling, go with *Webster's* and you'll never be wrong. Occasionally, two spellings will be given for the same word, and you'll have to make a choice. Here's where other style guides, such as the *U.S. Government Printing Office Style Manual*, may come in handy. Note that when *Webster's* says "or" (e.g., *disk or disc*), both forms are judged to have equal currency, and the fact that one of them must necessarily be first is not particularly significant. When *Webster's* says "also," the second spelling is significantly less current than the first.

□ SPLIT INFINITIVES

There is nothing inherently wrong with splitting an infinitive except that some educated readers may be offended by it. The careful writer will be aware of this sensibility and repair split infinitives wherever it is possible to do so without creating a strained sentence.

□ TELEGRAPH STYLE

Some writers have a habit of eliding the definite article when writing instructions ("Put tab through slot and tighten screw . . ."). In most instances, this does not enhance comprehension.

□ TITLES OF PUBLISHED WORKS

Titles of published works are set in italics.

□ USAGE

Two excellent books on usage are Fowler's *Dictionary of Modern English Usage* (2d edition, Oxford: Clarendon Press, 1965) and Theodore Bernstein's *Careful Writer* (New York: Atheneum, 1965). Fowler in particular will reward the reader with many hours of happy browsing. The following items represent some specific usage problems that are prevalent in technical writing.

Bolt. A *bolt* is a threaded fastener designed to be used with a nut. Most other externally threaded fasteners are properly called *screws*.

Compare. *Compare* takes the prepositions *to* and *with*. In technical writing, you will nearly always use *with*. You compare something to something else when you consider only the similarities ("Shall I compare thee to a summer's day?"). When you compare something with something, you consider both similarities and differences.

Data. *Data* and *media* are plural nouns. While *data* is gaining some acceptance as a singular noun, its use in the plural is never wrong ("These data are correct").

Employ. "How do I get this open?" "Use a screwdriver." That is the way that we talk. It is not the way that we write, however. For some unknown reason, we are likely to write that we "employ" or "utilize" a screwdriver. It is as if the ordinary term were somehow not good enough for writing. Like the constable who went to the scene of a disturbance but in court testifies that he "proceeded" there, we all have a formal and an informal vocabulary. The cause of clarity will be served, however, if we stick to the ordinary forms as much as possible.

Fluid. *Fluid* is not an exact synonym for liquid. Fluid includes liquids and gases. *Liquid* is the more precise term to use when referring to matter in the liquid state.

However. There is nothing intrinsically wrong with starting a sentence with a conjunction or a conjunctive adverb, but it is a little strange and when overdone strikes one as a nervous habit or tic. Starting a sentence with *However* is like starting a lecture with "Are there any questions?"

Output. The use of the noun *output* as a verb is now so firmly entrenched that there seems little hope of dislodging it. Nevertheless, it would be good if writers, as custodians of the language (nobody else gives a damn), were to remember that the verb form *put out* is still alive and write "The computer is required to put out 14 × 11 in. graphics to the printer" rather than "The computer is required to output . . ."

***That* and *Which*.** *That* and *which* are not interchangeable pronouns. *That* introduces a restrictive clause, *which* a nonrestrictive one. Failure to recognize this distinction is the solecism that perhaps more than any other serves to distinguish the amateur from the professional technical writer.

The relative pronouns *that* and *which* were undifferentiated as late as the sixteenth century. The distinction observed by most authorities, although of recent origin, is a useful one, especially in technical writing, as may be seen from the following examples:

> The purpose of the label is to identify the rotors which can be safely run in any given instrument.

The reader is uncertain whether this means that all rotors can be safely run in any given instrument or only the rotors so identified. A comma would make the meaning clear (but wrong):

> The purpose of the label is to identify the rotors, which can be safely run in any given instrument.

Substituting *that* for *which* removes all doubt:

> The purpose of the label is to identify the rotors that can be safely run in any given instrument.

Here is another example from the same technical manual:

> The L5 instruments have an imbalance detection device. This is not a convenience which lessens the importance of balanced loading.

Does this mean that the imbalance detector is a nuisance and therefore balanced loading is unimportant, or does it mean the opposite? A comma after *convenience* would clarify the matter:

> This is not a convenience, which lessens the importance of balanced loading.

This is obviously not what the writer intended. What he meant was:

> This is not a convenience that lessens the importance of balanced loading.

What is required of us is simply that we use the same unaffected style in writing that we use in speech. One would not likely say "This is the house which Jack built." Why write it? Somehow, with pen in hand, amateur writers (and a good many professionals) get the feeling that the simple, unaffected word is not lofty enough for the elegant prose that we are composing. The following paragraph and similar injunctions in *The Careful Writer* (Bernstein), *The Elements of Style* (Strunk and White), and Fowler's *Modern English Usage* emphasize the distinction between *that* and *which*.

> *That* is restrictive; a clause that it introduces is never set off with commas. *Which* is usually nonrestrictive. If we force ourselves to distinguish between these two, the structure of our sentences will become clearer to our readers.

Writers often complain that they are not able to tell whether a clause is restrictive or nonrestrictive. Perhaps the following will make the distinction clear.

Restrictive clauses identify the substantive under discussion to the exclusion of all others. In "This is the house that Jack built," the phrase *that Jack built* focuses the reader's attention on one particular house. Restrictive clauses are not parenthetical and hence do not require commas.

Nonrestrictive clauses perform a totally different function. They supply additional information about a substantive the identity of which is not in question. In the sentence "This house, which was built by Jack, is full of malt," the clause *which was built by Jack* supplies additional information about a house that, presumably, has already been identified. Nonrestrictive sentence elements are parenthetical and require commas.

□ WORD LIST

You may want to make a list of words that, for reason of spelling or usage, are likely to prove troublesome. Standard reference works are democratic, reporting variant usages with fine impartiality. The aim of your word list must be exactly the opposite: Where the dictionary says "disk or disc," your house style must choose one or the other. Another function of your word list will be to list usages that are peculiar to a particular industry or field.

Spelling. If a word has more than one spelling and the word crops up occasionally in your company's literature, by all means include it in your list. If there is only one correct spelling—and the word is as easily found in the dictionary as it would be in your list—leave it out.

Obsolescence. Occasionally, a term may appear as current in the dictionary but, as far as technical usage is concerned, be as dead as the dodo. I refer to such terms as *centigrade,* which have been declared obsolete by an international body. Such terms, with their acceptable alternatives, may be included in your word list.

Special Terms. Every discipline employs a special vocabulary of terms that do not appear in the average desk dictionary. Include these, too, if spelling or usage is a problem.

The following list, constructed according to the principles just established, may serve as a nucleus for a list of your own.

allen screw (use *hex socket screw*)
allen wrench (use *hex key*)
ampule
analogue
band-pass

bandwidth
butanol, 1-butanol (not *n-butanol*)
byproduct
canceled, canceling
cannot
caution—word used in instruction literature to call attention to possible hazards or unsafe practices that could lead to damage to the equipment. Cf. *note* and *warning*.
diagramed, diagraming
dialogue
disk
electronvolt
focused, focusing
formulas
hard copy (noun)
hard-copy (verb)
homogeneous (not *homogenous*)
indexes
indices (mathematics, crystallography)
isopropanol (use 2-*propanol*)
Kel-F (footnote to first use: 3M Company)
labeled, labeling
leveling
liquefy
maximal (adjective)
maximum (noun)
maximums (not *maxima*)
methyl Cellosolve (footnote to first use: Union Carbide Corp.)
micron (use micrometer)
Millipore (footnote to first use: Millipore Corp.)
minimal (adjective)
minimum (noun)
minimums (not minima)
note—word used in instructional literature to highlight essential information that can be safely ignored without hazard to equipment or personnel. Cf. *caution* and *warning*.

optimal (adjective)
optimum (noun)
path length
Phillips Screw
programed, programing
proved (not proven)
reexamine
sera (not *serums*)
setup (noun)
side chain
sizable
stand-alone
stud—a sometimes threaded fastener projecting from a machine
Teflon (footnote to first use: E. I. du Pont de Nemours & Company)
test tube
thermostated
toward
un-ionized
warning—word used in instructional literature to call attention to hazards that, if ignored, will lead to injury to personnel. Cf. *caution* and *note*.
X-ray (noun, verb, and adjective)

PART THREE

WRITING AND PUBLISHING

The following chapters take you step by step through the publishing process from conception to delivery of printed copies. As a writer, you will be intimately concerned with the research, organization, and manuscript phases of publishing. After that, your involvement diminishes but does not quite disappear. For this reason, some knowledge of the production and printing phases of publishing is essential.

The following paragraphs identify the things that must happen between generation of the first draft of a manual and delivery of printed copies.

First Draft. The first draft of a manual consists of all text and tables, and preliminary sketches for the illustrations, or notification of what is to come.

Keyboard. The text must be keyboarded on a word processor or typewritten prior to editing.

Review. Manuscripts should be circulated for technical review to the following departments or individuals.

Marketing. Manuscripts should be reviewed by the product manager to ensure compliance with corporate objectives and to ensure that they contain no information that can be used to the detriment of the company.

Engineering. Manuscripts should be reviewed by cognizant engineers to ensure completeness and technical accuracy.

Field Service. Manuscripts should be reviewed by the Field Service department to ensure the accuracy and appropriateness of any service information contained therein.

Product Safety Officer. Manuscripts should be reviewed by the product safety officer to ensure against inadvertent or implied recommendation of unsafe practices.

Legal Review. A copy of each manuscript should be submitted to the corporate legal officer. The concerns here are that the document must make no unsupportable claims for the product, that it must not offend another company, and that reasonable precautions must be taken to protect proprietary interests regarding trade names and the like.

Patents. Manuscripts should be reviewed to ensure that adequate notification is given regarding any features of the product that are protected by U.S. or foreign patents.

Revision. Manuscripts must be revised to incorporate the reviewers' comments. Any conflicts must be resolved at this time.

Illustrations. Line illustrations must be prepared from the writer's sketches as required. When time does not permit preparation of adequate line art, or where the view does not lend itself well to line illustration, photographs may be substituted for line art.

Copy Editing. All copy must be edited for consistency of spelling, punctuation, grammar, and the like. Edited manuscripts should be returned to the writer for final revision.

Final Draft. The final draft must indeed be final. It will have been reviewed and edited, and all the reviewers' and editor's comments will have been incorporated or resolved. The ribbon copy (not a facsimile) of the final draft will be submitted for production. This is the copy from which type will be set.

Production Editing. The final draft must be checked for subordination logic (sections and subheads) and agreement of sections and subheads with the table of contents. Levels of subheads must be designated. Cross-references to sections, figures, tables, and other literature; the list of illustrations; and the list of tables must be checked. Footnotes and displayed material must be marked for typesetting. Manuscript pages must be numbered. Placement of figures and tables must be indicated. The manuscript should be arranged for production as follows.

1. Title page
2. Table of contents
3. List of illustrations
4. List of tables
5. Specifications
6. Text
7. Tables
8. End matter
9. Illustrations
10. Legend copy for illustrations

A machine (xerographic) copy of the legend copy for the illustrations should be retained by the editor.

Book Design. A design concept should be prepared for each manual or each family of manuals.

Typesetting. The text, tables, figure captions, and callouts (lettering in illustrations) should be set in a style and size of type specified in the book design.

Galleys. Galley proofs must be returned to the writer to check for errors introduced in typesetting.

Pasteup. The corrected galleys must be combined with the graphics (artwork) to make up the camera-ready mechanicals for the printer.

Page Proofs. Page proofs must be returned to the editor to check for errors introduced during page makeup. When all is in readiness, the camera-ready copy is delivered to the printer.

Negatives. The printer must photograph the camera-ready copy and strip in the halftones to prepare the negatives from which the plates will be made.

Plates. The printer must then make the plates, which will be used to print the document.

Printing. The required number of copies must then be printed.

Collating. The printed pages must be collated following a printer's map or a dummy provided by the customer.

Binding. The collated pages must be trimmed and bound according to instructions provided by the customer.

Delivery. Bound copies must be delivered to the customer.

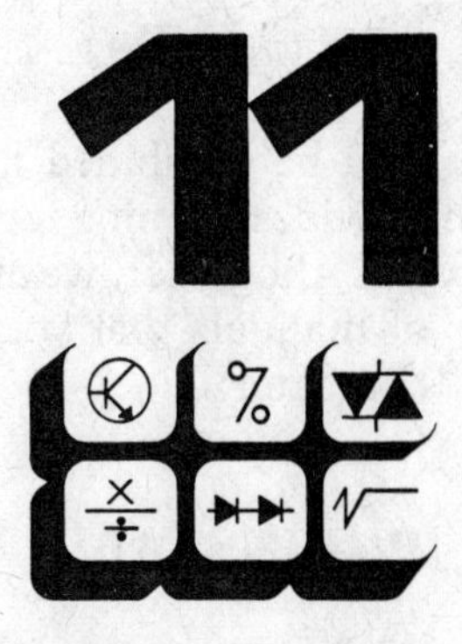

11 Technical Manuals

This chapter contains recommendations for the content, style, and format of technical manuals. The recommendations regarding content are based on the American National Standard for Preparation of Manuals for Installation Operation and Repair of Laboratory Instruments (ANSI C105.2–1972), available from the American National Standards Institute, 1430 Broadway, New York, New York 10018. This document in turn is based on the FDA "labeling" requirements for instruments used in *in vitro* diagnosis of human ailments. Despite some preoccupation with "adulteration" and "mislabeling," the FDA requirements make eminent good sense for the documentation of any sort of technical equipment.

The ANSI document recognizes the need for three types of documentation, based logically on the time, relative to the delivery of the product, when the documentation is needed. Preinstallation instructions are needed before the product is delivered, in order that the customer may prepare a site and the equipment may be installed and checked without delay. An operation and maintenance manual must be delivered with the product, and a field repair manual must be made available to the customer (at a "nominal" cost) sometime before the expiration of the warranty period. Of course, for a small system, all three documents may well be combined in one volume. In addition, if the company provides training for its customers, there may be a training manual. Altogether, we are talking about possibly four different kinds of manuals that you may be involved with. Each has a distinct character.

☐ PREINSTALLATION INSTRUCTIONS

Preinstallation instructions comprise the information that the customer must have prior to delivery of the equipment. Preinstallation instructions rarely run more than twelve or sixteen pages. Some things that the customer will need to know are the physical dimensions and weight of the equipment and the amount of clearance required all around for operation and maintenance, so that a suitable site can be prepared. Some questions that may be on the customer's mind are: Will it go through all the doorways between the receiving dock and the installation site? Can I bring it up in the elevator? Will the floor support the load?

The customer must know the utility requirements, the environmental requirements, the exhaust requirements, and so on, so that all may be in readiness when the manufacturer's representative arrives to install the system. Utility requirements must be specified so that power will be available at the appropriate potential, current, frequency, and phase. The power outlet receptacle must match the connector supplied with the equipment and be within reach of the equipment's power cord. If water is required, the flow rate, temperature, and purity must be specified. If venting is required, this, too, must be specified. The preinstallation instructions should also include a procedure for the initial check of the system.

Manufacturer's Policy Regarding Installation. The preinstallation instructions should stipulate the manufacturer's policy regarding unpacking and installation—that is, whether unpacking and installation are to be performed by the customer or by the manufacturer—and also the manufacturer's policy regarding moving the equipment after it is installed. If unpacking and installation are to be performed by the customer, provide appropriate instructions, including how to inspect for damage and the proper procedure to follow in the event that the product arrives in bad condition. Point out any hazards that may be associated with unpacking and installation.

The preinstallation instructions should list any tools or supplies that the customer must provide for use during installation and checkout. If any gases or chemicals are required, specify the purity and the recommended quantity.

Environmental Requirements. The preinstallation instructions must thoroughly specify the environmental limitations of the product. Environmental requirements may include temperature, humidity, exhaust, air conditioning, and the like.

Utilities. Specify the voltage, current, and frequency of electric power; any grounding, gas, water, vacuum, oxygen, or drains that may be required; and where and how the utilities are attached. Describe the connector required, and state the length of the power cord. Specify any vibration limits and any shielding requirements. Specify the heat load (Btu's per hour and watts—air conditioning may be required).

Limitations. Specify the temperature and humidity limits within which the product will perform to specifications.

Coupling. If the product is to be coupled to other equipment, specify the input or output impedance and provide troubleshooting information so that any coupling problems may be traced to either the product or the peripheral equipment.

□ OPERATION AND MAINTENANCE MANUALS

An operation and maintenance manual must be delivered with the product. There is no exception to this rule. The customer is entirely justified in withholding payment for the product pending receipt of a usable manual. (It may be in Xeroxed form.)

The operation and maintenance manual is the document that is usually referred to as "the manual." It tells the customer how to operate the equipment and what to do when it stops. Typically, this is all that the customer really wants to know about a product. Discuss limitations of the equipment and any precautions regarding cleaning in the operation and maintenance manual. Repeat the cleaning precautions in the field repair manual. The operation and maintenance manual should contain the following information (not necessarily in this order):

- use or function of the product
- installation instructions
- calibration
- operation
- hazards
- performance specifications
- troubleshooting
- maintenance
- principles of operation
- service information
- warranty

Use or Function of the Product. The manual should contain some mention of the use or function of the product. This elementary bit of intelligence is frequently overlooked. After all, the writer knows what the product is used for. Doesn't everybody? The answer is no. Frequently, a subtitle on the title page will satisfy this requirement. For example:

PATCHWORD
A Stringy Floppy Patch for Radio Shack's Scripsit

Installation Instructions. Include installation instructions, even if the product is to be installed by the manufacturer; the service representative may be glad for an opportunity to glance at the instructions if they are included in the manual. In any event, the installation instructions will come in handy if the customer decides to move the product after it has been installed.

Calibration. Include instructions for any calibrations that can be performed by the customer. Indicate a source for any calibration standards needed.

Operation. Instructions for operating the product should be placed near the front of the manual. Operating instructions should include a list of things that the operator should check before putting the system into operation. Describe clearly the method of putting the system into and out of operation. You may want to consider affixing this information to the instrument.

List a few things that the operator can do to make sure that the system is operating properly after it has been placed into operation. Include initial calibration methods and the means of adjustment. Explain the electromechanical controls and their ranges of operation. If some adjustments must be made before others, explain the sequence. Specify any adjustments that must be made with the equipment on, or with the equipment off. Include a technical summary that specifies the acceptable range of response of the system for various conditions.

Provide a precise, abbreviated summary of the operating instructions. This can be posted on the front panel of the equipment or on a readily accessible subpanel. This summary should include the emergency shutdown procedure.

Include a review of operating precautions ("Dos and Don'ts") at the end of the section on operation.

Hazards. Point out hazards and safety precautions at appropriate places in the manual, always before any step that exposes the operator to a potential hazard. The manual should discuss any electrical, fire, or explosion hazards and stipulate any exhaust requirements. Clearly describe the correct procedure for handling highly flammable or potentially explosive fuels such as acetylene, propane, hydrogen, or nitrous oxide. If oxygen or compressed air is used, the manual should state clearly that the user must install oil-free regulator valves if they are not supplied with the equipment. State the procedure for leak testing of all valves and lines in systems using compressed gas. Adequately describe any mechanical, chemical, optical, electrical, or radiation hazard; and issue proper warnings.

Performance Specifications. The operation and maintenance manual should specify the performance that may be expected of a system that is in good working order. Performance specifications should be placed prominently near the front of the manual.

Troubleshooting. Explain the adjustments that the operator can make to restore proper operation of the product before calling in the service representative. Include information that will enable the operator to determine whether it is the system or a utility that has failed. Include diagnostic steps that will enable the operator to say more than "It doesn't work" when talking to the service representative on the telephone. This will help the service representative decide what tools and supplies to bring along when making the service call. Include any troubleshooting procedures that the operator can perform, and a list of spare parts.

Troubleshooting charts, with boxheads labeled "Problem," "Probable cause," and "Remedy," are usually worthless. In the first place, the problems listed are seldom the problems that the customer has. In the second place, the formality of the chart leads the writer to the brink of idiocy, for example: "Problem: pilot light doesn't light; Probable cause: blown fuse; Remedy: replace the fuse."

Maintenance. Include any maintenance procedures that can be done by the operator. Do not include in operator manuals any information that may encourage inexpert tinkering. Include a schedule for preventive maintenance.

Principles of Operation. Explain how the product works. Remember that many customers will not have a service contract. Institutional users of complex systems in particular frequently find it to their advantage to employ a maintenance man to keep all of their equipment going. In such instances, it frequently happens that the person who is called in to service the product has never seen one in working order. A description of how the equipment is supposed to function is then more valuable than a troubleshooting chart.

Describe the scientific principles on which the system operates in sufficient detail for a person unfamiliar with the equipment to understand how it works. This should include an operational description, functional diagrams, and descriptions of the relation of each subassembly to the system.

Service Information. Tell the operator how to get help when he or she needs it. After the operating instructions, the most important part of an operation and maintenance manual is the part that tells the operator what to do if the equipment fails. Explain fully the manufacturer's repair policy, including a description of repairs that can be done by the user. This section should describe the adjustments or repairs that can be done locally and those that must be done at a regional repair center. It should list the locations of the regional centers and specify any repairs that should not be attempted by the user and that would void the warranty. Training programs, workshops, telephone service centers, special tools, preventive maintenance procedures and schedules, and maintenance records are appropriate topics for this section of the manual.

Warranty. The warranty statement is a quasi-legal document. Don't mess with the wording, syntax, or punctuation; it's not your affair. The warranty statement must be part of the manual, not a separate document, so that the customer will not be able to maintain that he or she didn't receive it. Copies of service contracts may be included. State the period for which repair parts will be available.

□ FIELD REPAIR MANUAL

The field repair manual isn't needed until the end of the warranty period. For a new product, this gives the engineers a chance to get a little experience with the product before committing themselves to specific service recommendations. The field repair manual contains all of the schematics, wiring diagrams, and circuit descriptions needed by the field service engineer to repair a malfunctioning system. Appropriate contents for the field repair manual may be summarized as follows.

- physical descriptions of the subassemblies
- diagrams
- electrical and electronic descriptions
- mechanical descriptions
- field calibration and adjustment procedures
- troubleshooting

Physical Descriptions of the Subassemblies. Describe the various subassemblies in enough detail so that they can be identified by the person who is providing service. Describe the operation of each subassembly, any operating or calibration adjustments, and the effects of such adjustments.

Diagrams. The field repair manual should include external views of the product showing controls, indicators, and connectors. Describe how to remove covers and shields. Include views of the equipment with the covers removed, wiring diagrams, and detailed diagrams as required. External views should include callouts (lettering on illustrations) for controls, indicators, connectors, and the like. It may be necessary to diagram more than one level or one view of the equipment. Include a diagram of the wiring of the system, including color codes and any descriptions needed for clarity. Include enlarged diagrams of any part of the system that is not clearly defined by the diagrams just described.

Wiring Diagrams. Wiring diagrams are used by service personnel to trace the electrical connections between one subassembly and another.

Schematic and Logic Diagrams. The field repair manual should include overall schematic or logic diagrams, parts lists, illustrations showing the physical location of parts with test points and voltage waveforms, and a list of spare electrical and electronic parts. Point out any hazards.

Schematic and logic diagrams are used by service personnel to trace circuits at the circuit-board level. Such diagrams must be legible, comprehensive, and easily read without magnification. Include parts lists of electrical and electronic components with their values; and list acceptable substitutes, if any, with their commercial designations. Such lists should include the component part number for ease of ordering.

Describe or illustrate the physical location of each part. Specify test points providing access to important measurable voltages, resistances, and waveforms. Specify voltage waveforms to be expected at various test terminals. Include a list of suggested spare electrical parts to be kept on hand, especially any that are hard to get or that may fail without warning. Emphasize any potential electrical or mechanical hazard.

Timing Diagrams. Timing diagrams show the time relationship between signals in different parts of a circuit and help the service representative to understand how the circuit works.

Waveforms. Waveform diagrams show the service representative what voltage waveforms may be expected at various points in a circuit and are thus an aid to troubleshooting.

Component Location Diagrams. Component location diagrams, including circuit board layouts, enable the service representative to locate specific components and test points.

Parts Lists. Parts lists list the parts for each subassembly with their descriptions and part numbers for ease of ordering.

Electrical and Electronic Descriptions. Describe the electrical and electronic circuits, keying the descriptions to the appropriate schematic or logic diagrams.

Mechanical Descriptions. Describe any mechanical functions that are not obvious upon inspection. The field repair manual should include a parts list and a list of spare parts. For optical systems, include a description and diagram of the optical pathway and the means of adjustment. For precision equipment, include descriptions and diagrams in the field repair manual. Include a list of mechanical parts, acceptable substitutes with commercial designations, if any, and a source of supply. Include the part numbers and descriptions for ease in ordering. Include a suggested list of spare mechanical parts to be kept on hand because of frequent failure or procurement difficulties.

Field Calibration and Adjustment Procedures. Include any calibration procedures that are not included in the operation and maintenance manual. The field repair manual should list any tools required for calibration and adjustment and describe the locations of access ports and the methods to be used. Designate any factory-recommended instruments, tools, and standards necessary for precise calibration and adjustment. Describe the precise location of ports providing access to adjustment screws. Adjustment procedures should include mention of any potential hazard to personnel or to the instrument.

Troubleshooting. Include a procedure for rapidly locating malfunctioning components. Such a procedure will usually involve opening any feedback loops, because a malfunction within a closed loop will give an abnormal indication at every point in the loop. Rapid troubleshooting requires successive division of the circuit until the malfunctioning component is isolated. This section may also include troubleshooting hints, if experience indicates that certain faults and symptoms are likely to be associated.

□ TRAINING MANUALS

Customer training courses will likely use the operation and maintenance manual as a textbook and require in addition a laboratory syllabus or training manual. The training manual will be of limited usefulness after the training course is completed.

☐ STYLE AND FORMAT

A published document has three elements: content, style, and format. *Content* refers to the information contained in a document; it is that which remains invariant whether the communication is oral or written and whether the language used is English, Chinese, or Swahili. The writer is responsible for the content of the document.

The word *style* is used in two senses. There is mechanical style, which includes spelling, punctuation, grammar, usage, and the use of italics, capital letters, bold type, hyphens, and the like to enhance the effectiveness of written communication—the sort of thing that constitutes Part II of this book. Then there is literary style—the quality that distinguishes the prose of one writer from that of another. It is style in the first sense—mechanical style—that concerns us here. Style is invariant whether the document is handwritten, typed, or printed. The editor is responsible for the style of the document.

The elements of style—spelling, grammar, and punctuation—are the ground rules of prose. Take away the rules, and what do you have? Poetry, perhaps, but certainly not technical communication. Is there then no place for creativity in instructional literature? Indeed, there is.

The term *format* refers to the aesthetic or subjective side of publishing. Good format enhances the message by permitting the reader's attention to focus on the message without distractions from the means of presenting the message. *Format* refers to the appearance of the document, in contrast to its style and information content. The size of the page, the width of margins, the typeface and size, the length of the lines and the number of lines per page, whether the right-hand margins are justified or ragged, whether the text is in two columns or one, and many other details are matters of format. The art director is responsible for the format. This is not to say that the writer has no voice in the matter. Indeed, if the writer has ideas on how a document should look, he or she has a responsibility to make these ideas known to the art director in a preproduction conference. Ultimately, however, it is the art director, not the writer, who is responsible—who takes the praise for a good-looking publication and gets the blame for an ugly one.

Style is conservative. Material written 200 years ago will appear quite modern if printed in a modern format, since the rules of spelling, punctuation, and grammar change very slowly. A moment's reflection will reveal why this is so. Style communicates on the conscious level; it is therefore essential that the rules of spelling, punctuation, and grammar remain relatively constant so as not to interfere with the message. Format, on the other hand, communicates on the unconscious level. Today's manuals would look quite old-fashioned if printed in a format only twenty years old. This is because advances are continually being made in typography and layout.

Page Size. It's an 8½ × 11 world. Bookcases, file cabinets, desk drawers, and "in" baskets are all scaled to accommodate documents that measure 8½ × 11 inches. If for some reason the art director sees fit to depart from this format, he or she must be prepared to defend such a decision.

Binding. The length of the document is an important consideration in determining the type of binding. Major system manuals running 100 pages or more are usually encased in loose-leaf three-ring binders. Loose-leaf documents have the advantage of lying flat when open. Shorter documents may be saddle stitched. Very short documents (one to six pages) are usually not bound.

Typeface. Thousands of typefaces have been designed, although few typesetters have more than a dozen or so. Not all typefaces are appropriate for every document. For instance, a typeface that is appropriate for wedding announcements would not be appropriate for technical manuals, and vice versa.

Typefaces that are appropriate for technical manuals may be divided into two classes: serif and sans serif. Serif typestyles are somewhat more legible than sans serif type, possibly because the eye is guided by the serifs from one letter to the next. Sans serif typestyles convey a modern, progressive image. The text of this book is set in a serif type, and the chapter and part titles are in sans serif.

Type Size, Leading, and Line Length. Selection of a type size for the body copy is a compromise between legibility and economy. *Leading*, a term held over from the days of hot type composition, when strips of lead were used between lines of type, refers to the amount of space between lines. The amount of leading will vary with line length, longer lines requiring more leading because the eye wearies of trying to find the beginning of the next line. Ten points on eleven or twelve (that is, one or two points of leading) by nineteen picas is about ideal. This allows two columns of text on an 8½ × 11-inch page. Notes, cautions, warnings, and indexes are sometimes set in type that is one point smaller than the body copy and with a point less leading. Tables and figure captions are frequently set in a typeface that is different from that used for the text.

Justification. Justified copy (copy in which the white space between words is adjusted so that all lines of type will be the same length) is slightly harder to read than unjustified (ragged right) copy. Typesetters justify copy because that is the way that Gutenberg did it, and Gutenberg did it because that is the way that the medieval monks did it, to save paper and because they thought that God liked it that way. Unjustified copy is widely used in advertising, but it is common for the text of manuals to be justified.

Word-Processor Printouts. It is often difficult for technical publications to keep up with engineering design changes, and it is sometimes necessary to forego typesetting and use the word-processor printout directly for camera-ready copy. This is particularly true in software documentation.

Management of White Space. The key to setting legible information is the intelligent management of white space in a manner that permits the eye to grasp quickly what is to be read and then read it with a minimum of interference. In the days of hot type, typesetters knew that the most expensive thing in typesetting was the setting of white space. Whereas lines of type were rapidly set by the Linotype operator, the white space had to be filled in by hand with strips of lead and wood "furniture" in all the non-printing areas. With the advent of what is commonly called "cold type," white space suddenly became very cheap; you simply put the black where you wanted it and let the white take care of itself. Now, with computer composition, white space is becoming expensive again. After the text is edited, it is necessary to go over the copy with an eye for form, adjusting the white space for optimal appearance and legibility.

Headlines. Each section heading should be on a new page and placed about a third of the way from the top. Since word processors lack the variety of typefaces and sizes that are available to the typesetter, the various levels of headlines must be distinguished from one another by other means. If your word processor can produce bold type, this is very useful in distinguishing headlines. Otherwise, horizontal rules may be used to good effect. L E T T E R S P A C I N G may be used to enhance the importance of a headline. Leave two blank lines before and one blank line after first- and second-order headlines. Third-order headlines may be run in with the text. Underlining is an editing mark that should not be used in printed documents.

Text. If your word processor is capable of two-column operation, the text may be single spaced. The text must be double spaced if it runs all the way across the page. Leave an extra blank line between paragraphs.

Notes, Cautions, and Warnings. Notes, cautions, and warnings may be set off from the rest of the text by indenting the copy a few spaces from the left.

The remaining chapters of this book provide a step-by-step description of the preparation of a typical operation and maintenance manual for a major system, from project inception to the delivery of printed copies.

12 The First Draft

This chapter describes the development of a rough draft for a fairly routine technical writing assignment: an operator's manual for a complicated piece of equipment. Less ambitious projects will not make use of all the topics discussed here; but in describing the draft of a major manual, we shall touch on most of the problems that a beginning technical writer is likely to encounter on a first assignment.

□ RESEARCH AND ORGANIZATION

A technical writer is like a reporter. As a technical writer, you must interview cognizant persons and read the background material before you can begin to write. A good place to start is with the marketing manager, who is the person responsible for the profit earned by the product. Although the marketing manager may not be the best informed on the technical details, he or she is in the best position to give you an overview of the product and a general orientation regarding what the company expects in the way of a manual.

The next person that you should talk to is the project engineer. He or she is the person who knows more about the product than anyone else, because he or she designed it. You will depend on him or her for a thorough technical explanation of how the product works. Other valuable resources are the field service engineer—for information on troubleshooting and maintenance—and the customer training department for information on operation.

A very valuable source of information is literature that has already been written: old manuals, marketing literature, and the like. Engineering drawings are another valuable source of information. Get the silkscreen drawings for instrument panels so that you will be able to refer to controls and indicators exactly as they are placarded on the panels. You will also need assembly and part drawings, schematics, block diagrams, wiring diagrams, parts lists, specification drawings, and test procedures, as appropriate for the level of documentation that you will be writing.

Attend the product development meetings. The earlier you start attending these meetings, the better. Writing the manual will be a breeze if you have been involved in the project from the start.

There is no substitute for hands-on experience. When you are able to get your hands on the product and actually operate it, you will find that many of the things you have written, while perhaps technically accurate, can be restated to better effect. Try to schedule some time when the product can be made available for your use. This is not always easy to arrange; there are many demands for time on a new product: from marketing, from field service, from applications specialists, from training personnel, and so on. In addition, there is pressure to get the product out the door and billed to the customer.

Just as a house is not built without a plan, so is a book not written without an outline; ignore this advice at your peril. Each hour spent in outlining is worth several spent in writing. In time this will become second nature to you, and you will scarcely write a note to the milkman without outlining it first.

Let us say now that you have acquired a large amount of knowledge concerning the system that you are going to write about. You know in a general way how to operate it and what to do when it stops. You will want to set up major divisions (called "sections"). Your first attempt may look something like this:

Section 1. Operation
Section 2. Troubleshooting
Section 3. Maintenance
Section 4. Principles of Operation
Section 5. Theory

Don't worry too much about the order; it's easy to move things around later. Your next step would be to think up some major subdivisions for each section:

Section 1. Operation
  Controls and Indicators
  System Startup and Shutdown
  Calibration
  (Etc.)

Another approach could be to divide the system into subassemblies and devote a section to each subassembly. Both approaches have their merits; and no matter which strategy you decide upon, there will be people who will wish that you had done it differently. You can't please everybody. The subassembly approach is attractive if the same subassemblies are used in different products. If you use the subassembly approach, your outline might look something like this:

Section 1. Subassembly A
  Operation
  Troubleshooting
  Maintenance
  Principles of Operation
  Theory

Section 2. Subassembly B
  Operation
  (Etc.)

Either way, your outline must remain flexible. If you revise it continually as you write, it eventually will become the table of contents for your book.

☐ GETTING STARTED

After you have done your research and written an outline, you are ready to begin to Write. You sit down at your work station, and you type: "Section 1. Introduction." Then, if you are like most writers, a curious phenomenon sets in. It is something like paralysis of the cerebrum. Few things are perfect in this world; but at the moment, your mind is very close to being a perfect and absolute blank. There is no cure for this. Cigarettes are worse than useless; they cause cancer.

There is no cure for writer's block, but there may be a way around it: *Don't start with Section 1.* Start anywhere else. Start with material that is fresh in your mind and fairly straightforward. If you start with the last section and work your way forward, you may find that by the time that you reach Section 1 you will be so familiar with the material that Section 1 will virtually write itself.

A manuscript differs from a printed document in several important respects: It is double spaced, it is printed on one side of the paper only, and it lacks the variety of type styles and sizes that are available to the typesetter. For these reasons, it is not practical to try to indicate in a manuscript how the finished document will look. Indeed, such attempts can be misleading and lead ultimately to sullenness and disillusionment on the part of the reviewers. This is not to say that the writer should not take an interest in the layout and in all aspects of the publishing process, but the time for that is after the manuscript has been through the review process and is being readied for production.

☐ MANUSCRIPT CONTENTS

A typical technical manual may contain text, tables, and illustrations. These disparate elements should be brought together in the first draft to form a coherent whole. If one element—for instance, the illustrations—is left until later, you may find that you will have considerable rewriting to do. The first draft of a book-length manuscript should include the following:

- title page
- table of contents
- any other preliminaries
- text
- tables
- list of references (if any)
- any other end matter
- illustrations
- legend copy for the illustrations

Your manuscript will be reviewed; and all of these things, which are part of your book, should be reviewed at one time.

Title Page. Your manuscript is not complete if it does not include copy for the title page. The title page should carry the proprietary name and the generic name of the product. The proprietary name of the product (e.g., *Ford*) should never be used alone; this weakens the proprietary claim to the name. Always use the common name with the proprietary name (e.g., *Ford Motor Car*).

Table of Contents. Although a table of contents is not actually necessary for the first draft of your manuscript, it is helpful to give reviewers an idea of the structure of the manual. The table of contents should list the section headings and at least the first-order subheads.

Specifications. Operators' manuals should include a table of specifications, including operating characteristics, environmental factors, and utility requirements.

Text. The text of an operators' manual should include the information described in Chapter 11 for operation and maintenance manuals.

□ COPY PREPARATION

Your first draft should be produced on a word processor or typewriter. Manuscripts submitted for review should reflect the same standards of professionalism expected of manuscript submitted for publication. Generous margins (about 20 mm) should be left at top, bottom, right, and left. Start each manuscript or each section of a major manual about halfway down the page. Indicate clearly where each section begins. Don't leave the editor and designer in the dark regarding which headlines represent major divisions of your manual.

If you use a typewriter, type on one side of the paper only. All copy must be double spaced. This applies to text, tables, titles, captions, cautions, notes, warnings—everything. No single-spaced copy should appear anywhere in the manuscript. The only exception to this rule is entire paragraphs lifted verbatim from previously published documents. *Double spaced* means a full blank line between all typed lines. Single-spaced lines longer than 70 or 80 mm are difficult to read; the eye tires of finding the beginning of the next line after scanning a long line of text. If this were not so, newspapers would have gone to 360-mm columns years ago. Single-spaced copy is also impossible to edit clearly and is difficult for the typesetter to follow. Use a slash (only one per term) when a quotient is typed in line.

Word Processing. Intelligent management of white space can contribute greatly to the readability of a manuscript. Start each section of the manuscript about twelve lines from the top of the page. Insert two blank lines before first- and second-order headlines and one blank line before each paragraph. Indent displayed material (notes, cautions, and warnings) about five spaces, and leave extra space (one blank line) before and after.

Insert one blank line before each item in a series of numbered steps, including the first; and insert one blank line after the last item. Align runover lines with the first word in the first line of text. For example:

1. Now is the time for all good men to come to the aid of their party.

Use the same style for lists, if items in the list run longer than one line. Format all copy for double-spaced printout. (This means that at some point before you print out your file you should give your word processor the format command that will cause the printout to be double spaced.)

Headlines. Headlines are part of the graphics, not part of the text; but since they are so closely related to content, it is customary in technical writing for the writer, not the editor, to write the copy for the headlines. Differentiate headlines by their placement on the page; for instance, first-level headlines may be centered, second-level headlines may be typed flush left, and third-level headlines may be run in with the text. If you have need of a fourth-level headline, it could be that you haven't done your outlining right.

Do not use underscoring, because it will have to be removed by the editor in the final draft. Similarly, headlines that are written in ALL CAPITALS are harder for the editor to mark for typesetting than are headlines that are in Capitals and Lower Case.

Body Copy. When a hyphen is used at the end of a line, the typesetter cannot tell if the hyphen is to be retained when the word is set in type. Use a double hyphen (=) if the hyphen is to be retained.

trouble- shooting	troubleshooting
double= dealing	double-dealing

Displayed Material. Set displayed material (notes, cautions, warnings, and equations) off from the rest of the text by indenting about five spaces and leaving extra space above and below.

There is no clear agreement on just what it is that serves to differentiate a "note" from a "caution" and a "caution" from a "warning." One protocol that is sometimes seen is as follows: A note is a message that may safely be ignored without hazard to either equipment or personnel. A caution is a message that calls the reader's attention to unsafe practices or conditions that, if ignored, could result in damage to the equipment. A warning calls attention to unsafe practices or conditions that, if ignored, could result in injury to persons. This is a protocol that is simple to apply but that could lead to apparent contradictions. For instance, consider a situation in which you want to caution an operator regarding a condition that could result in loss of a sample. The foregoing protocol would require that the appropriate message be labeled "Note"; but if the operator had spent several years in working up the sample, he or she might well consider the loss of the sample of greater consequence than possible damage to the equipment or perhaps even the risk of personal injury. At the other extreme, consider the use of a substance that may produce a slight rash in certain susceptible individuals. The foregoing protocol requires that a message to this effect be labeled "Warning." A more sophisticated protocol would take into account the degree of risk involved and also the gravity of the consequences.

Glossaries. When preparing glossaries, lists of abbreviations, and the like, type the entries flush left, in alphabetical order. Do not use a period unless the period is part of the entry. Leave four spaces between the longest entry and the definition. Align all definitions. Use flush-and-hang indentation. If any of the definitions consists of more than one sentence, use periods after all the definitions; otherwise, do not use periods.

Bibliographies. When preparing bibliographies, type each entry flush left. Use flush-and-hang indentation. Bibliographies—and all parts of the manuscript—must be double spaced. Leave two blank lines between items. Write authors' names last name first. If there are several works by the same author, use a dash (three or four hyphens) for each work after the first. If a period follows the author's name in the first item, a period follows the dash, also.

□ TABLES

Use tables when tabular presentation of the data is more efficient than text. Use text when text is more efficient than tables. Tables should be on pages separate from the text. (A short table may be run in with the text.) Horizontal rules can be produced on the word processor or typewriter. If columns are properly aligned, vertical rules should not be necessary.

Parts of a Table. A table consists of at least five parts (see Figure 24): table number, title, boxhead, body, and stub. Footnotes to a table are part of the table, not part of the text, and are

Figure 24. Parts of a table: (1) table number, (2) title, (3) boxhead, (4) body, (5) footnote, (6) stub. A headnote, if used, would appear in the space following the title, above the double rule.

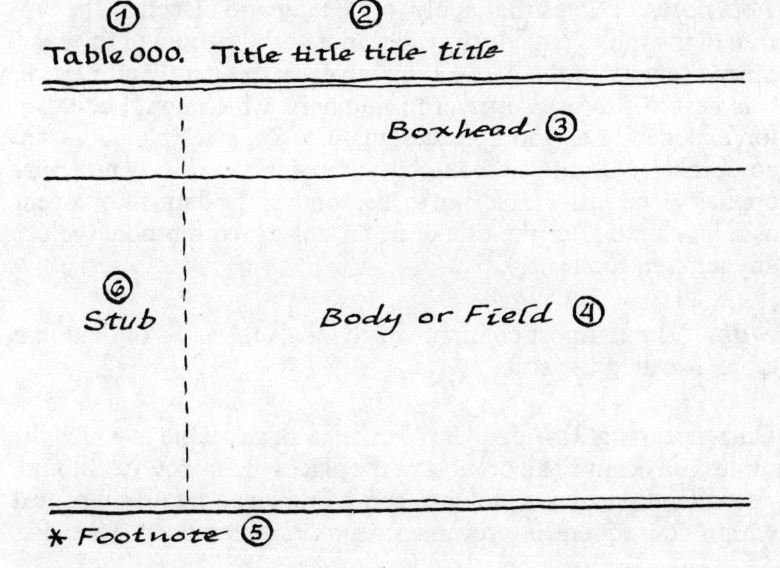

placed at the foot of the table (outside the rules) rather than at the foot of the page.

Table Number. Tables and illustrations cannot always be placed in the book exactly where you would like them. For this reason, it is necessary to number the tables and refer to them by number in the text. The table number and title are traditionally placed at the head of the table.

Title. The title of the table follows the table number, either on the same line or on the next line.

Boxhead. A table consists of rows and columns. Each column has a heading that applies to the entire column. The column headings are referred to collectively as the *boxhead*.

Body. The field of information below the boxhead is called the body of the table.

Footnotes. Notes that apply to the table are placed at the bottom of the table (outside the rules), not at the bottom of the page. Since the data in the body of the table are frequently numerical, it is best not to use superscript numbers, which could be taken for exponents, as footnote designators. Superscript letters are acceptable. Do not use footnote symbols in the title. A note that applies to the entire table, not to just some of the data, may appear as a headnote after the title or as an unlettered footnote (before any lettered footnotes).

Stub. The leftmost column, which labels the rows of data in a table, is called the stub.

Constructing Tables. Be suspicious of any table in which the same data occur in all or most of the places in any row or column. The offending row or column can be replaced by a line of text with a consequent improvement in overall efficiency. Likewise, be suspicious of a table that contains a great many blanks. Experiment with various arrangements of the data, exchanging rows for columns, and so on, in order to find the most economical presentation, bearing in mind the shape of the standard 8½ × 11–in. page format. In some very simple tables it is even possible to turn the table inside out, substituting body copy for the boxhead or stub.

Consider the following hypothetical data: A company manufactures two models of Turboencabulators. Each is available in two colors and at two prices. Let us call them Model X and Model Y. Say that a red Model X costs $1000 and a blue Model X costs $2000. A red Model Y costs $3000, and a blue Model Y costs $4000. You probably wouldn't want to tabulate such simple data, but suppose that you did. You might start out like this:

Table A.
Turboencabulator Colors.

	$1000	*$2000*	*$3000*	*$4000*
Model X	red	—	blue	—
Model Y	—	red	—	blue

Now, suppose that you would prefer a more vertical presentation for use in a two-column format. Try interchanging the boxhead and the stub.

Table B.
Turboencabulator Colors.

	Model X	*Model Y*
$1000	red	—
$2000	—	red
$3000	blue	—
$4000	—	blue

That's better, but you may be bothered by the fact that half the spaces in the body of the table are blank. Try interchanging the stub and body. Note that the title of the table changes so as to describe what is in the field or body of the table.

Table C.
Turboencabulator Prices.

	red	*blue*
Model X	$1000	$2000
Model Y	$3000	$4000

Now that's more like it—a very compact table with no blanks. Note that this table contains all of the information that is in Table A, but in more economical form. The purpose of this exercise is to demonstrate that you don't have to be stuck with someone else's format, if you can see a way to improve on it.

□ ILLUSTRATIONS

Illustrations are classed as either line art or halftones, depending on the process used to prepare them for printing. Line art consists of black lines and areas on a white background, which can be photographed along with the type to prepare plates for offset printing. Halftones also consist of black areas, but the black areas are rather small, produced by photographing a continuous-tone original such as a photograph, pencil sketch, or watercolor through a fine screen to produce a negative consisting of tiny dots of various sizes, which when printed give the effect of the continuous-tone original. In general, in technical writing *halftone* means "photograph."

When well done, line art is capable of considerable elegance and is preferable to halftones in a class-A technical manual. Line illustrations are expensive in comparison with photographs; but if the illustrations have to be revised, the cost of revising line art is usually less than that of bringing a photographer in for another session. In many instances, this may be an important consideration when one is weighing the economics of line art versus photographs.

Unless you are unusually gifted, you will not be drawing the illustrations yourself. Nevertheless, the illustrations are your responsibility and you should be thinking of them as you write your first draft, or you may have considerable rewriting to do later on. Any changes in the illustrations are likely to affect your text.

When reviewers read "See Figure 1," they would like to have some idea of what Figure 1 looks like. This is a good reason for including sketches with your first draft.

The better your sketches are, the better the finished art will be; technical illustrators are not known as "line straighteners" for nothing. One of the most valuable courses of instruction that you can take in preparation for a career in technical writing is a course in mechanical drawing. You will learn not only how to read engineering drawings but also how to make orthographic and isometric sketches. (Orthographic projections show three views of an object: top, front, and side. Isometric drawing chooses a point of view from which all three aspects—top, front, and side—can be seen at once.)

The Photo Session. One of the many things that a technical writer has to do in addition to writing is to supervise the photography session when pictures are taken for use in a manual. Things will go smoother if the writer takes the time to prepare a shot list beforehand, visualizing camera and subject positions and any necessary disassembly of the product, so as to make the most efficient use of the photographer's time. If a model is needed for some of the shots, these shots should be grouped together so that the model's time is used to best advantage. Give a copy of the shot list to the project engineer and to anyone else whose cooperation is needed for the photo session.

Availability of the product will be a determining factor in setting a date for the photo session. You will not be able to photograph the product before it is built nor after it has been shipped. This is critical, because there is frequently great pressure to ship the first unit as soon as it is built and bill the customer for it, which reduces the amount of time available for photography. You may need the intervention of the marketing manager in order to arrange time for your photo session.

Marketing will need photos, too, for publicity. Resist any pressure to share a photography session with marketing; their needs are not the same as yours. They need color shots; you need black and white. They need glamour shots; you need documentation. You need a product that is authentic in every detail; they couldn't care less. You need to watch your budget; they have money to burn. You need to see inside the product; they are satisfied with an authentic-looking mockup. Let them have their photo session and get out of the way.

Choose a photographer who specializes in industrial photography. It is the only way that you can be sure that horizontal lines in the photograph will be parallel to the lines of type and vertical lines will be perpendicular when the picture is printed on a page with text. Photographing shiny machinery is not necessarily more difficult than photographing nature, but it is different.

You will have to arrange a place to work, with enough room for the photographer and his or her equipment to move around the product as required for the various views and to move away for an overall view. Talk to the photographer about his or her requirements.

Schedule the session for a time when the project engineer can be available to answer any questions that may come up and to verify that the configuration of the product actually represents the way that it will be built. Make certain that the positions of any controls showing in the photographs make sense. It is best to have the engineer set each control to some reasonable position.

The art director should be present to pass judgment on the artistic merit of each shot and to be alert to any background problems that may cause difficulties in preparing the photographs for reproduction.

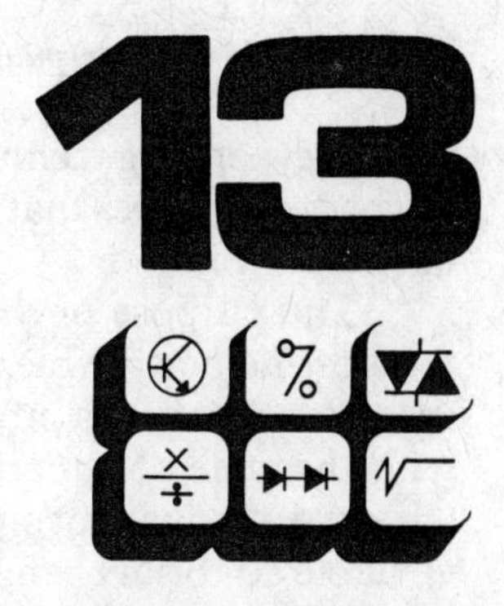

Review and Revision

The terms *review*, *edit*, and *proofread* are used almost interchangeably outside the publishing field; however, each has a precise meaning in publishing.

The term *review*, as used in technical and scientific publishing, means critical evaluation of a manuscript for accuracy and completeness of information content. Reviewing is performed by persons selected for their expertise in various technical and scientific disciplines.

Like reviewing, editing is performed on the manuscript rather than on the galleys; but unlike reviewing, the goal of editing is evaluation and improvement of the efficacy of communication rather than the accuracy of the information content. Editing is performed by persons selected for their competence in communication skills.

Proofreading means examining typeset copy to check for errors introduced during the typesetting or pasteup processes. Proofreading is performed on galleys and page proofs, never on manuscripts.

You are responsible for the technical content of your manual. There is no way that you can avoid this responsibility. Since you cannot, of your own knowledge, vouch for every statement contained therein, it is essential that the manuscript be circulated for review among technical experts. These will, for the most part, be the very sources that were consulted during the research phase of the project.

The purpose of the review cycle is to obtain for the writer the best help available. The reviewer's responsibility transcends that of an editor or a proofreader. Whereas the editor is concerned with the style of a manuscript, the reviewer must be concerned with the manuscript's contents. Reviewers have at times made valuable editorial changes and have even written entire manuals, but such contributions are peripheral to the main consideration, which is the critical evaluation of manuscripts to ensure their technical accuracy. If this entails rewriting or editing, so be it, but the writer must not expect the reviewer to do any writing, rewriting, editing, or proofreading. Proofreading, especially, is wasted on all but the proofs.

It is sometimes necessary to explain to the reviewers exactly what it is that is expected of them. This can be done in a note or cover letter attached to the manuscript. See Figure 25 for a sample of such a cover letter.

The proper time to review a document is before type is set. The principal impediment to efficient publishing is the necessity of doing over that which has already been done. This is seen at its worst in setting type that has already been set and in tearing up art that has already been pasted down. Efficient operation demands that technical review of publications must be completed in the manuscript phase. Review of documents after type has been set invites costly revisions and delays.

□ REVIEWERS

Reviewers should be selected so as to elicit response from the widest possible range of interests and areas of expertise. There is little to be gained by soliciting comments from reviewers whose interests and areas of expertise are identical. In general, solicit comments from the following departments or their counterparts

```
Note to Reviewers

This draft has been written and edited by a staff with a flaming
passion for accuracy. NEVERTHELESS, it is inevitable in a work
of this magnitude that some errors will have crept in. We are
not speaking here of errors of spelling, punctuation, and grammar
(although Lord knows these abound in even the best-edited of
texts) but errors of FACT, which can do our image in the market-
place scant good. The purpose of the technical review is to
reduce such errors to a minimum.

Your comments on style are welcome and will be read with
interest. Please bear in mind, however, that as the printer's
deadline approaches, we work with one eye on the CLOCK instead of
the calendar. If all of your comments do not get incorporated into
this edition of the manual, we beg your indulgence. Each comment
must be carefully evaluated and perhaps debated. Outright
errors, whether of fact or usage, will of course be corrected.
We will not knowingly print errors in a new manual. Other
improvements must of necessity take second place.

The same considerations apply to QUESTIONS posed by reviewers.
Legitimate questions deserve answers--and there is never time at
this stage in the editorial process to do the necessary research--
so our plea to you is: DON'T GIVE US QUESTIONS WHEN IT'S ANSWERS
THAT WE NEED. The odds are overwhelmingly against the answers'
getting into the manual unless you give them to us.
```

Figure 25. Sample cover letter.

in your organization: marketing, engineering, field service, customer training, and legal. The following paragraphs describe some typical individuals and organizations whose opinions might be solicited.

The Product-Line Manager. The product-line manager is responsible for the profit or loss generated by the product. This person has a vital interest in the timeliness and appropriateness of the documentation. In particular, the product-line manager will be alert to any statements in the documentation that could be used by the competition to your company's disadvantage.

The Project Engineer. The person who designed the product is the one best qualified to pass judgment on the technical accuracy of your manuscript's content. After the product-line manager, this is your most valuable reviewer.

Field Service. The field service engineer gets out into the field and talks to customers. This is the person who gets it in the neck if your manual is wrong. Pay close attention to his or her comments. Adequate documentation can help to reduce the number of service calls to which the field service engineer must respond. The field service engineer accordingly has a legitimate interest in reviewing manuals to ensure that they meet field service requirements.

Training. If your company has a customer training department, your manual will likely be used as a textbook. Make sure that it meets their needs. Operation and maintenance manuals that meet the needs of the training department will help to reduce duplication of effort, to the ultimate benefit of the company.

Legal. Send a copy to the legal department. You may not get a response, but at least you will have done the right thing. Your concern here is twofold: (1) that you have not made any unsupportable claims for the product and (2) that you have not inadvertently or by implication said anything actionable about anybody else's product. Another concern is that you have taken reasonable precautions to protect the company's proprietary interests in trade names and the like.

Safety. Your manuscript should be scrutinized to make sure that you have not inadvertently or by implication recommended any unsafe practices. In recent years, courts have tended increasingly to find in favor of the plaintiff in cases where product safety is involved. If your company has a safety officer, he or she might be the appropriate reviewer for this concern.

Patents. Someone should review the manuscript to ensure that adequate notice is served regarding patented features of the product.

☐ SUGGESTIONS FOR REVIEWERS

In order that the review process consume as little of the reviewers' time as possible, some suggestions such as the following for marking a manuscript may be offered.

1. Delete irrelevant material by drawing a line through the material to be deleted. Whole paragraphs can be quickly deleted in this way.
2. Indicate in the margin where the text needs to be expanded.
3. Indicate corrections by crossing out the incorrect word or expression and writing the correction directly above it. All material is double spaced for this purpose.
4. Draw a line around material that is out of place and indicate in the margin where you think the material belongs.
5. Don't give us questions—it's answers that we need.

☐ INCORPORATING THE REVIEWERS' COMMENTS

Reviewers' comments can be divided into two categories: comments on content and comments on style. All have to be read and carefully evaluated. Although you are the style expert, you are not perfect; and some things may have escaped your notice. Some reviewers may know much more about English spelling, punctuation, and grammar than you. For the most part, however, the stylistic comments will convey personal preferences at best and be downright mistaken at worst. Comments of this sort will take up a disproportionate amount of your time, but you dare not ignore them wholesale. Try to evaluate such comments honestly, incorporate the ones that improve the text, and ignore the rest.

Content is another matter. Here, the reviewer is the expert, not you. Evaluate such comments carefully, seek elaboration if you do not understand them, and revise the text as required.

☐ RESOLVING DIFFERENCES

A writer went to his supervisor with a problem. "I have a conflict," he said. "Marketing claims 1 percent accuracy for this product, and the engineer says it is only 2 percent. What shall I do?" "No problem," said his supervisor. "You simply tell the truth."

This is good advice. The writer alone is responsible for the content of the document; you cannot shift this responsibility to the reviewers. In the event of conflicting information, listen carefully to both sides and make up your own mind. Since you will have to bear the blame for any errors anyway, you may as well tell it as it is. Be discreet. It is frequently possible to omit from your manuscript any passages that flatly contradict existing product literature, rather than compound the offense. Bear in mind that the customer has already bought the product; you do not have to sell it again.

If you are not able to resolve the conflicts on your own, then call a meeting of the interested parties and thrash the matter out. This should be done as a last resort. You have only to add up the salaries of the high-priced talent for the duration of the meeting to realize what such meetings cost the company.

□ PREPARING FOR PRODUCTION

When you have incorporated or resolved all of the reviewers' comments and are thoroughly sick of the document and wish more than anything to be rid of it, then go over it carefully one more time. It is this "one more time" that more than anything else serves to distinguish the professional from the amateur writer.

You're not out of the woods yet. When you are satisfied with your manuscript and believe it to be the finest thing that you have ever done, there is a surprising amount of work yet to do to get it ready for production. You have to write the copy for the title page and the table of contents and maybe a list of illustrations and a list of tables. You have to decide where to place the product specifications (and perhaps even write them). You've got to double check the cross-references to section numbers, figure numbers, table numbers, and the like. If the document is typewritten, there may be some retyping to do. Are the pages numbered? You must tell the production people where the figures go and where to put the tables.

Manuscript Contents. A complete manuscript for a major manual should include copy for a title page, a table of contents, and specifications. If the illustrations are of interest in themselves, without reference to the text, then a list of illustrations is required. If the illustrations are unintelligible without reference to the text, then a list of illustrations is not needed. Entries in the list of illustrations should be identical to the actual captions, except that a long caption may be curtailed for inclusion in the list of illustrations. The same considerations apply to a list of tables.

Corrections. Corrections should be made in ink, always above the manuscript line—never below it. Delete words and phrases with a horizontal line through the offending material. One or two handwritten corrections per page are quite acceptable, but a page with many corrections should be redone. Number all the text pages in the manuscript.

Cut and paste or use transparent mending tape to rearrange material. If pages are added in the middle of a manuscript, use lower-case letters to identify the added pages; for instance, *11a, 11b,* and *11c* identify pages added between pages 11 and 12. In this instance, write "11a follows" at the bottom of page 11 and "12 follows" at the bottom of page 11c. The following correction methods are unacceptable:

- corrections written on the backs of pages
- corrections written up or down the margins
- slips attached to the pages
- additions to the bottoms of pages, folded up
- instructions to the typesetter on separate pages
- corrections written in pencil

Correlation. Cross-references are a fertile source of error. Read the entire manuscript, checking all cross-references to section numbers, figure numbers, table numbers, and other literature. If the manuscript contains footnotes, read it again, checking the footnotes. If the manuscript contains a list of references, read it one more time, checking the citations in the text against the list of references. Retype any pages requiring many or extensive corrections. Number the pages.

Marking the Manuscript. Special instructions to the typesetter regarding type changes should be marked directly on the manuscript.

Display Type. Use a single underline to specify italics and a double underline to specify small capitals. Use a wavy underline to specify bold type.

Headlines. Indicate the level of each headline. For instance, you might use an *A* for the first level, *B* for the second level, and *C* for the third level. Circle the notations, so that they will not be set. Do not underline headlines, as the book designer will specify the type to be used for each level and the editor may have to delete your underlining.

Displayed Matter. Draw a vertical line in the margin adjacent to notes, cautions, and warnings to alert the production editor to the necessity for a change in type size or face.

Symbols. Identify handwritten symbols (Greek letters, etc.) with a note in the margin.

Illustrations. Collect your illustrations in a separate folder. Paste small illustrations on $8^1/2 \times 11$–in. paper. Provide 8×10 prints for all photographs. Print out the legend copy for the illustrations and include it in the folder with the illustrations. Provide the artist with a list of callouts.

Placement of Figures and Tables. Using a pencil, indicate clearly in the margin where each figure and table belongs. Circle the notations so the typesetter will know that they are not corrections. The production people will place each item as near to where it belongs as possible, but they will not place it earlier in the document than the position indicated. Mark each illustration with its figure number and the number of the manuscript page where it belongs. Circle the notation. Do the same for the tables. Identify each photograph, using a pencil very lightly on the back. Indicate the top of the picture so that it won't be printed upside down in the manual.

List of Illustrations. If the manuscript includes a list of illustrations or a list of tables, check these against the figure captions and titles. The wording in each instance should be identical, except that a long caption or title may be shortened for inclusion in the list of illustrations or tables.

Arrangement of the Manuscript. The arrangement of the contents of a manuscript that is submitted for editing is generally not the same as the arrangement of the contents of the printed book. The front matter and the text will be in the usual order, but the tables should be printed out on separate pages and grouped after all the text matter. The sketches for the illustrations should be in a separate folder, and the legend copy for the illustrations should be printed out separately and included in the folder with the illustrations. (Xeroxed copies may be placed on the pages with the illustrations.) Table titles should be on the pages with the tables.

Writer's Checklist. Manuscripts should be edited prior to typesetting. Edited manuscripts are then usually returned to the writer for final corrections. The following checklist lists some of the things that will be checked during editing. Considerable time may be saved if these items are checked by the writer prior to submitting the manuscript for typesetting.

Front Matter

- Title page
- Table of contents
- List of illustrations
- List of tables
- Specifications

Correlation

- Section numbers
- Figure numbers
- Table numbers
- References to other literature
- Footnotes

Miscellaneous

- Retyping (if required)
- Pagination
- Placement of figures and tables

14 Production and Printing

In the early days of printing, manuscripts went directly from the author to the printer; and the printer took the responsibility for editing—such as it was—as well as book design and setting type and reading proof. Nowadays, the printer gets mechanicals, which he photographs to make the plates from which the book is printed. The typesetting, proofreading, and page layout are all done before the printer ever sees it. These intermediate steps are called "production."

When the manuscript goes into production, your involvement diminishes and you can turn your attention to other things. You will have to respond to the editor's queries, and you will have to proofread the galleys and page proofs and maybe write an index; but for the most part, the pressure is off. Nevertheless, until the day that the mechanicals finally go to the printer, there will be questions that only you can answer.

To understand the production process, it is necessary to know a little about the printing process, to which production is prologue. Basically, there are three ways in which ink can be transferred from metal plates to paper.

(1) The ink can be applied to raised surfaces, which are then pressed against the paper. This process, called letterpress, was

the way that Gutenberg's press worked and is the way that most printing was done until quite recently. Rubber stamps work this way.

In high-speed letterpresses, the ink must be carefully formulated to remain wet long enough to transfer to the paper but then dry immediately thereafter. Occasionally, if the ink doesn't dry quickly enough, some of the ink will transfer from the freshly printed page to the back of the next sheet in the stack. This phenomenon, called offset, so undesirable in letterpress work, is the basis of another kind of printing called offset lithography.

(2) The ink can be applied to an engraved plate and then scraped off by a "doctor blade" so that ink remains only in the pits and grooves of the engraving. If the ink is thin and the paper absorbent, the ink will be sucked from the grooves by capillary action. This process, called gravure (or rotogravure, with rotary presses), is sometimes used to print magazine supplements for Sunday newspapers.

(3) The ink can be transferred from the plate to a rubber roller, which is then pressed against the paper to make an image. This two-stage process, called offset printing, is virtually the only method used for printing technical manuals today.

The plates used in offset printing transfer the ink to the rubber roller by yet another process called lithography (literally, stone writing). Although no stone is used in modern lithography, the process derives from a method used in the early part of this century by artists to produce multiple prints. A stone was ground flat and very smooth, and the artist used a wax crayon to draw a picture on the stone. The stone was then flooded with water and inked with a roller. The ink, which has an oil base, would not adhere to the wet places but would adhere to the waxy image. Paper pressed to the stone would then be imprinted with a replica of the original design.

Offset lithography makes use of the same principle. The plate bears an image that is neither raised as in letterpress printing nor engraved as in rotogravure but simply treated chemically so as to attract oil and repel water. The image is not drawn with a crayon but produced photographically. For this reason, the process is sometimes called photolithography. As in other photographic processes, each plate is printed ("burned") from a negative, which is exposed in an enormous camera. The masters that are photographed to make the plates are known as mechanicals.

□ EDITING

Editing is the liaison between the writer and production. It consists of two separate functions: copy editing and production editing. Copy editing is a service to the writer; production editing is a service to the production staff. An editor must work with the writer to try to ensure that what the writer has said is indeed what will actually be understood by the reader. The editor also must work with the production staff to ensure that the writer's words get proper exposure—that nothing gets lost in the reduction of words into type.

Copy Editing. Manuscripts are edited to improve the effectiveness of communication between the writer and his readers. Important goals are the elimination of ambiguities and the improvement of sentence structure.

The manuscript that is submitted for editing must indeed be final. It will have been through all required reviews, all changes and corrections will have been incorporated, the writer will have checked it thoroughly, and the cognizant marketing manager will have approved it. It is something that the writer and the reviewers can live with for now and (we may hope) for months to come.

After all reviewers' comments have been incorporated or resolved, the ribbon copy of the manuscript will be submitted to the editor. (You should keep a copy for yourself.) The ribbon copy is marked by the editor and becomes the copy from which type is set.

Edited manuscripts are returned to the writer for correction and approval before type is set. At this time, you should examine the final edited manuscript again very carefully to be certain that all your facts are correct, that you are satisfied with the copy, and that you have answered all of the editor's questions. The next copy that you see will be a proof; changes cannot be easily made once type is set.

All final corrections should be on this final manuscript; it is the document against which proofs will be read. If you are not absolutely certain that the final manuscript is final, you should not release it for production until you *are* certain. To be sure, making the last careful pass through the document is tedious after having lived with it for so many months; but overcoming such tedium and doing a good job is the mark of a professional.

Last-minute corrections are inevitable, and there is only one way to make handwritten corrections in a manuscript: cross out the incorrect word or expression and write the correct expression directly above it. (This is another reason for double spacing.) Any of the following irregularities are generally considered unprofessional:

- single-spaced copy
- overtyped corrections
- corrections written in the margin
- corrections written under the line
- extensive handwritten corrections

Writer as Editor. It often happens that, in a shop that is too small to afford the services of an editor, the editorial duties devolve upon the writer. If that is your lot, you must then take off your reporter's snap-brim fedora and don the green editorial eyeshade to take yet another look at the manuscript that you are by now so thoroughly sick of. If you have prepared your manuscript well, according to the principles laid down in Chapter 14, then this should not take long. Some writers use green ink when they are playing editor. When editing your own or someone else's work, perform the following steps.

1. Double check the section headings and subheads to make sure that the subordination makes sense and that the headlines agree with the table of contents.

2. Check all cross-references to sections, figures, tables, and other literature, and also the list of illustrations and the list of tables.

If a standard format has not been established, you may need to specify the type for your manual.

3. Specify the typeface that is to be used.
4. Specify the type size, leading, and line length to be used for the body copy. (Ten points on eleven or twelve by nineteen picas is ideal.)
5. Specify whether the body copy is to be justified or ragged right.
6. Specify the type size, weight, and style (capitals, capitals and lower case, italic, etc.) for first-, second-, and third-order headlines.
7. Specify the type size and leading, white space requirements, and alignment of displayed material.
8. Specify the type for boxheads (top and lower levels), body, and notes of tables. Specify also the thickness of the rules.

Displayed material (notes, cautions, warnings, and equations) must be marked in some way so that the typesetter will be alerted to the necessity of changing the type size, weight, or measure (if so specified). One way of doing this is by drawing a vertical red line in the margin.

The level of each subhead must be indicated with a character (e.g., *A*, *B*, or *C*). Circle the character so that it will not be mistaken for a correction and set.

9. Mark footnotes for type change. One way to do this is with a vertical blue line in the margin.
10. Specify the amount to indent flush-and-hang runovers; for example, "Indent flush-and-hang runovers 2 ems."
11. Specify material that is to be set in outline style: for instance, numbered steps. If outline style is specified, runover lines will be set flush with the text following the number. For example,
    a. This text is set outline style. Note that runover lines are set flush with the beginning of the text above (not flush left).
12. Make sure that the writer's name appears at the top of each page.
13. Mark clearly the beginning and end of each section (for example, "begin 1," "end 1").

14. If there is an appendix, write "appendix follows" on the last page of the text.
15. Indicate in the margins where the figures and tables should be placed. Transfer these notes to the galleys after type has been set.
16. If you haven't already done so, number your pages.

Most typesetters know that letters used algebraically are to be set in italic type, but it is a sensible precaution to underscore them anyway. Identify any handwritten symbols (e.g., "Greek mu") the first time each is used. Use carets to clarify superscripts and subscripts. For example,

$$y^{2}$$

You may find it convenient to prepare a checklist to remind yourself of all the things that you must do to prepare your copy for production:

- Redline displayed material
- Mark subheads
- Blueline footnotes
- Specify flush-and-hang material
- Specify outline style
- Identify and place material
  - Writer's surname
  - Section begin and end
  - "Appendix follows"
  - Tables and figures
- Pagination
- Mathematical copy

☐ PREPRODUCTION CONFERENCE

At this time, the writer and the art director should have a preproduction conference to discuss format. The artist will prepare a design concept for the manual or family of manuals. The text, tables, figure captions, and callouts will be set in a style and size of type specified in the book design.

Binding. Common methods of binding for technical manuals are saddle stitching, perfect binding, comb binding, and loose-leaf binding. Other methods exist, but these are the most common. Very short manuals (one to six pages) are not bound.

Saddle Stitching. The use of wire staples to hold a book together is called stitching. If the sheets are stapled in the center, then folded and trimmed, the process is called saddle stitching. Saddle stitching is appropriate for manuals up to about fifty pages long.

Perfect Binding. A binding method that is perfect for paperback books and telephone directories can't be all bad. Perfect binding is "perfect" in the sense that the perfect tense in grammar is perfect; it refers to an action that is completed (which is to say that no further operations are required after binding). In perfect binding, the pages are trimmed first, then glued together along the binding edge. Perfect binding is an economical binding method for any size of manual.

Comb Binding. In comb binding, the pages are held together by a plastic comb with teeth that go through rectangular perforations in the margin of the paper. Comb-bound manuals will lie flat when open. This is an advantage.

Ring Binding. Three-ring loose-leaf binders are ideal for major manuals 200 pages long or longer. They lie flat when open, are easily updated, and look good on the shelf.

Covers. Manuals longer than six or eight pages should have some kind of cover. If the cover stock is the same as the paper on which the manual is printed, the cover is described as a "self cover." Manuals longer than about sixteen pages will generally have a heavier cover. The writer must provide copy for the cover, and for the spine, if any.

□ TYPESETTING

Modern typesetting is a far cry from the processes that were used in the days of metal type. What is produced is not really type at all but a marvelously sharp, clear photograph of what the impression would look like if metal type had been used. It's all done with optics. The output from the phototypesetter is in the form of long galleys that resemble the paper used for black-and-white photographic prints; and indeed, that is what they are. The galleys are cut apart and pasted down with the line art to prepare the mechanicals. Galley proofs will be returned to the writer to check for errors introduced in typesetting.

□ THE DEADLINE

After type has been set, the manuscript is "dead." No mark of any kind should ever be made on a dead manuscript. The deadline is the day and hour on which the manuscript is officially dead. Changes after the deadline should be limited to the correction of errors, whether typos or bad information.

It is a commonplace in the publishing industry that type once set is never reset except to correct errors. The reasons for this are threefold: (1) typesetting is very expensive by comparison with word processing, (2) extensive changes after type is set may delay publication as much as six weeks for new layout and pasteup, and (3) such changes inevitably provide opportunity for the introduction of new errors.

□ PROOFREADING

All copies of the document after the deadline are called proofs. It is standard publishing practice to make two sets: galley proofs and page proofs. Galley proofs are to be checked for errors introduced in typesetting; page proofs are to be checked for errors introduced during page makeup.

Proofs are for checking, not changing. Changes in proof are discouraged since they delay production and offer exceptional opportunity for the introduction of new errors. The proof stage is not the time to learn new facts about the product. Bear in mind that if the product is viable, the current document is not the last word that will be written on the subject. There will be ample opportunity to incorporate new information in future revisions of the manual. Most alterations are avoidable because they should have been made in the manuscript before production began.

Of course, some changes are inevitable. One must correct errors, whether typos or errors of fact. A typeset document looks different from one that is typewritten, and some minor changes in format or syntax may be necessary. Also, previously unnoticed inconsistencies have a way of cropping up once a document is typeset. Careful craftsmanship in writing and editing should keep such inconsistencies to a minimum.

Word Division. The computer program has not yet been written that can figure out where to break a word at the end of a line of text. Nor is such a program likely to be written. Instead, a new generation of readers will come to accept random word division, just as the present generation has come to accept junk food and background music.

End-of-line word division is the typesetter's worry, not the writer's; but, particularly in this day of computerized typesetting, words are likely to be broken any-old-where; and if you care about such things, you may want to call the typesetter's attention to particularly bad breaks. First, however, let us discuss some of the good places to break a word.

Good Breaks. American practice is to divide words between syllables, according to pronunciation (*pro-nun-ci-a-tion*). This sounds simple until you try to decide where, exactly, one syllable ends and the next begins. When in doubt, consult *Webster's New*

Collegiate Dictionary (not the unabridged). Note, however, that there are acceptable alternative end-of-line divisions just as there are acceptable variant spellings and pronunciations. Many of the word divisions given in *Webster's* would be difficult to defend on the basis of any arbitrary rules. For reasons of space, and to avoid confusion, the dictionary gives only one division where several may be acceptable. In general, no one is likely to question any reasonable word division so long as the five rules listed under "Bad Breaks" (following) are rigorously observed and a conscientious effort is made to avoid the "So-so Breaks."

- The suffix *-ing* may nearly always be separated from the root (*separat-ing*).
- Doubled consonants provide a dandy place to divide words (*suf-fix*). For that matter, whenever any two consonants occur between vowels (*con-sonants*), you may divide between the consonants.
- Compound words (*in-to*) may be divided into their original components.
- You may divide after a prefix (*pre-fix*).
- If all else fails, divide after vowels, unless this produces an unpronounceable fragment (*fragme-nt*).

So-so Breaks. The following breaks are not good, but neither are they strictly forbidden.

- Try to avoid dividing words that don't look right when divided (*wo-men*).
- Try to avoid carrying two letters to the next line. Bear in mind that the space occupied by one of those letters will be taken by a hyphen, so the most that you will gain will be one space.
- If a compound term already has a hyphen in it, don't compound the confusion by adding another (*one-syl-lable*).
- Try not to divide figures (*100*
  000).
- Try not to divide divisional marks—(1), (a)—from what follows.

Bad Breaks. And finally, a few proscriptions:

- Never divide a one-syllable word (*wo-rd*).
- Never carry to the next line a final syllable in which the *l* sound is the only audible vowel (*audi-ble*, *sylla-ble*).

- Never divide on a single letter (*e-nough*).
- Never divide between the initials of a proper name (*E. B. White*).
- Never separate figures from units, if the units are abbreviated (*50 ml*).

Dashes. The editor and the typesetter should know the difference between an en dash and a hyphen. (En dashes are used mainly between figures to indicate a range of values, as in 1880–1884.)

Equations. A long equation should be broken on an operational sign, preferably an equals sign; and the operational sign should be carried to the next line. Stacked equations should be aligned on the equals signs. For example,

$$\begin{aligned} \mathrm{s} &= (1.295 \times 10^{-9})\,(6.33 \times 10^{-4}) \\ &= 8.20 \times 10^{-13} \\ &= 8.20\ \mathrm{S} \end{aligned}$$

A displayed equation that is typed thus in manuscript:

$$\mathrm{t} = \mathrm{k/s}$$

must be set like this in the galley:

$$t = \frac{k}{s}.$$

Equations should be centered under the copy.

Multiplication Sign. The letter *x* (or *X*) should not be used for a multiplication sign in typeset copy.

Proofreaders' Marks. The following proofreaders' marks are standard throughout the publishing industry.

Inserts

period . . . . ⊙

comma . . . . ⁁,

colon	/:
semicolon	/;
question mark	?
hyphen	=/*
apostrophe	∨
en dash	1/N
em dash	1/M
space	#
lead	ld>
virgule	shill
Superior	∨
Inferior	∧
Parens	(/)
Brackets	[/]
Indentation	
one em	□
two ems	□□
paragraph	¶
no paragraph	no ¶
Transpositions	
of letters (strike out and insert)	
of words	tr
Spell out	sp
Italic	ital
Boldface	b.f.
Small caps	s.c.
Roman type	rom
Caps	caps

*Since the proofreaders' mark for a hyphen resembles an equals sign, perhaps we should explain how to insert a missing equals sign. To insert an equals sign, make the insertion at the appropriate place in the line and identify the sign with the word *equals* in quotes and circled so that the word *equals* will not be set.

Caps and small caps	c.& s.c.
Lower case	l.c.
Wrong font	w.f.
Close up	⁀
Delete	ℓ
Delete and close up	ℓ̂
Move	
right	]
left	[
up	⊓
down	⊔
Align	
vertically	‖
horizontally	=
Center	
horizontally	][
vertically	⊔⊓
Equalize space	eq. #
Let it stand	stet
Broken letter	⊗
Carry over to next line	run over
Carry back to preceding line	run back
Something omitted	out, see copy

□ GRAPHICS

Illustrations for the manual are collectively known as graphics and classed as either line art or continuous tone (photographs). The artist will prepare inked drawings from the writer's sketches. It is the writer's responsibility to review the finished art and verify its accuracy.

☐LAYOUT AND PASTEUP

After the galleys are corrected, the artist prepares a layout of the manual, combining the text with the illustrations. This usually goes pretty quickly. In the layout process, the text is regarded as so many square inches of gray background against which are displayed the notes, cautions, warnings, equations, illustrations, and tables. The artist does not have any discretion regarding the placement of the notes, cautions, warnings, and equations; but the placement of the figures and tables will be determined in part by artistic considerations and in part by the rigors of space. (A figure or table cannot be split but must be all on one page.) When the artist has a layout that works, the next step is to paste up the actual mechanicals that will be photographed to make the plates from which the book will be printed.

☐PAGE PROOFS

Page proofs will be returned to the author to check for errors introduced in page makeup. Look for omissions and duplications. (The text has already been proofread as galleys, so you don't have to do that again.)

The text and line art will appear as they will be in the printed document, but the photographs will likely be represented by black "windows." This is because continuous-tone illustrations require an additional step to prepare them for printing. The film that the printer uses to make the negatives is formulated so as not to register shades of gray—everything must be either black or white. Continuous-tone illustrations are photographed through a screen, which breaks the image up into a pattern of tiny dots. The size of each dot is determined by the local shade of gray.

The continuous-tone originals are pasted onto separate pieces of illustration board and sent along with the mechanicals to the printer. The printer photographs the line art (including the text) and the photographs separately, then strips the screened negatives (now called halftones) into the clear windows (black in the page proofs) of the line-art negative. It is good practice to request blueprints of the stripped negatives before the plates are made. This is your only chance to check the halftones to make sure that none has been flopped or switched or is upside down.

□ COLLATING

A large manual may sometimes include smaller documents that have been printed separately. If this is the situation, collating instructions must be given to the printer.

□ DEADLINE

Finally, the day comes that you have been waiting for. About fourteen working days (on the average) after the mechanicals have been delivered to the printer, you should receive printed copies. Skim off a few before the rest are put into stock. You will need at least four copies for your own use. One should be stamped "History Copy," and one should be stamped "Work Copy." Keep the other two for cannibalizing when you make the next revision or a new manual for the next generation of the product. If you are pleased with the way that the manual turned out, you may want a fifth copy for your portfolio.

Keep the history copy as a permanent record of exactly what was said in a given document at a particular time. You are the one who will be consulted when questions arise. A manual is a quasi-legal document, and courts have been increasingly likely to find in favor of the plaintiff in cases where documentation is in question. (It is possible for not only the company but the writer to be sued, in cases where personal injury or property damage results from inadequate documentation.)

Keep the work copy to revise as errors come to light or engineering changes develop. It will be the basis for your next revision. Also, distribute a few copies to the people who have helped you in the development of the manuscript—the product manager, the project engineer, and the others. Take the typesetter and the artist to lunch.

Glossary

ACCELERATION Any change in the speed or direction of motion.

BINDING A term used to describe the way that the pages of a book are held together.

BOOK DESIGN The aesthetic considerations regarding the format of a book.

CALLOUT Lettering on an illustration.

CAMERA READY The final stage of production prior to printing.

COLLATION The arrangement of the parts of a printed book.

COMB BINDING A type of binding that uses a plastic comb to hold the pages together.

CONFIGURATION PREFIX An italic prefix (*cis-* or *trans-*) attached to the name of an organic compound to specify which of two possible stereoisomers is intended.

CONTENT The information contained in a document; that aspect of a document which remains invariant regardless of style or format (or even language).

COPY EDITING Editing a document for style.

DEADLINE The day and hour on which a manuscript becomes

dead. No changes of any kind should ever be made on a dead manuscript.

DEFORM To change the shape of a material body.

DERIVATIVE A word formed by adding a prefix or suffix to another word.

DRAFT A stage of manuscript development. A manuscript may be revised many times before it is set in type. Each revision is called a draft.

ECHO (verb) To send a duplicate character to another file, as when a CPU, upon receiving a character from the keyboard, sends a copy of that character to the CRT.

EDIT To revise a manuscript for the purpose of improving the efficacy of communication between the writer and his readers; see also REVIEW and PROOFREAD.

FERRULE A short tube or bushing used to make a hermetic seal between a tube and another part.

FINAL DRAFT A manuscript that has been reviewed and revised and with which the writer is satisfied.

FORMAT (noun) The visual appearance of a document; cf CONTENT and STYLE.

FORMAT (verb) To enter print format codes into a word-processed document.

GALLEY Text that has been typeset but has not been made up into pages.

GRAPHICS All printed images exclusive of text, including but not limited to illustrations, headlines, rules, page numbers, design elements, cover designs, running heads, and spine art.

HALFTONE A printed reproduction of a continuous-tone illustration, such as a photograph, pencil rendering, or watercolor; also, an illustration intended for such treatment; cf LINE ART.

HEADNOTE A note similar to a footnote but used at the head of a table rather than at the foot.

INTERNATIONAL SYSTEM OF UNITS (SI) The metric system.

ISOMETRIC A style of mechanical drawing in which all three orthographic views (top, front, and side) are represented in one view. Length and angular measurements are to the same scale in all three dimensions; cf ORTHOGRAPHIC.

KEYBOARD (verb) To enter text into a word processor or phototypesetter.

KILOPASCAL One thousand pascals.

LAYOUT Designer's plan for page makeup; the arrangement of text and graphics on a page.

LEADING The white space between lines of type.

LINE ART Illustrations that can be reproduced without screening; cf HALFTONE.

LOCANT An italic letter (*S, N, O,* or *P*) used in the name of an organic compound for the purpose of specifying the point of attachment of a radical.

MACHINE (verb) To shape metal by drilling, milling, turning, boring, or other machine process.

MACHINE COPY A duplicate copy of a manuscript or illustration produced by xerography or other facsimile process.

MACROCOSM The universe that is visible to the naked eye.

MALT Grain that has been allowed to germinate.

MANUSCRIPT Any draft of a document prior to typesetting.

MICROCOSM The microscopic and submicroscopic structure of reality, imperceptible to the unaided senses.

NOMINAL In name only. A nominal dimension names the item; *e.g.,* there is nothing "one inch" about "one-inch pipe."

ORDINARY WORK Work not requiring extraordinary precision.

ORTHOGRAPHIC A style of mechanical drawing in which three views (top, front, and side) are used to represent an object; cf ISOMETRIC.

PAGE PROOF A copy of a document after it has been made up into pages but prior to printing, used to check for errors introduced during page makeup.

PANEL LEGEND Lettering on the panel of an instrument.

PARAMETER A quality (such as length) to which a numerical value may be assigned.

PASTEUP Mechanical assembly of the visual elements of a document, following the designer's layout; the final stage of production prior to printing.

PERIPHERAL Electronic equipment, frequently of another manufacture, used in conjunction with a computer.

PICA A typesetter's measure approximately equal to 1/6 inch.

POINT A unit of type size approximately equal to 1/72 inch. Twelve points = one pica.

POTENTIAL Electromotive force, "voltage."

PRINTING Printing, as described in this book, includes the preparation of negatives and plates from camera-ready art.

PRODUCTION The process of converting a manuscript into camera-ready copy prior to printing; all stages of publication between but not including writing and printing. Production comprises typesetting, layout, illustration, and pasteup.

PROOF Any copy of a document after the deadline and prior to printing. Galley proofs are checked for errors introduced during typesetting. Page proofs are checked for errors introduced during page makeup.

PROOFREAD To check proofs for errors introduced during production; cf EDIT and REVIEW.

PUBLICATION Any printed document designed for dissemination outside the organization.

PUBLISH To make public.

PUBLISHING Public dissemination of printed matter.

REFEREE Someone who reviews a manuscript prior to publication; cf REVIEWER.

REVIEW To scrutinize a manuscript for information content; cf EDIT and PROOFREAD.

REVIEWER Someone who reviews a book after publication; cf REFEREE.

REVISE To prepare a second or subsequent draft of a manuscript.

RIBBON COPY Hard copy of a document produced by a typewriter or a printer connected to a word processor; cf MACHINE COPY.

SCOPE Oscilloscope.

SCRIPT A typeface resembling handwritten letters.

SERIF A short finish stroke usually made perpendicular to the ends of the strokes of printed characters. This book is set in serif type.

SKETCH Writer's conception of an illustration, submitted to an artist for execution.

SPIRAL BINDING A method of binding that uses a helical wire to hold the pages together.

STRESS Force tending to change the shape of a material body.

STYLE As used in this book, the word *style* means the conventions such as spelling, punctuation, and grammar that are consistent throughout a document or series of documents and are constant regardless of whether the document is in magnetic, hard-copy, or printed form; the rules of spelling, punctuation, grammar, usage, and the like that remain invariant through the stages of publication, as distinguished from *format*. This is not to be confused with a writer's literary style, which is beyond the scope of this book; cf CONTENT and FORMAT.

SUBATOMIC Relating to distances that are small by comparison with the diameter of an atom.

TABULA RASA A blank tablet.

TEXT Words assembled according to grammatical principles. A document may consist of text, tables, and illustrations.

TRADITIONAL MEASUREMENT Nonmetric measurements (feet and inches; pounds and ounces; pounds, shillings, and pence).

TYPESETTING Visual presentation of written material characterized by a greater variety of type styles and sizes than are available on a word processor or typewriter.

UNIT MODIFIER Compound adjective.

USAGE The rules governing the use of words, abbreviations, and symbols.

WHEATSTONE BRIDGE A device for measuring electrical resistance.

VACUUM Pressure less than ambient.

Index